Scared Witless: Prophets and Profits of Climate Doom

Print ISBN 978-1-941071-21-2
eBook ISBN 978-1-941071-22-9

Other books by Larry Bell:

Climate of Corruption:
Politics and Power Behind the Global Warming Hoax

Cosmic Musings: Contemplating Life Beyond Self

STAIRWAY PRESS—SEATTLE

Cover Design by Chris Benson
www.bensoncreative.com

STAIRWAY PRESS

www.StairwayPress.com
1500A East College Way #554
Mount Vernon, WA 98273 USA

Dr. Fred S. Singer

Dedication

THIS BOOK IS dedicated most particularly to Dr. Fred Singer, along with tens of thousands of other very highly principled researchers and writers who are passionately committed to high standards of climate science integrity essential to restore badly compromised public trust.

Courageous willingness to challenge ideological orthodoxy and politically driven policy agendas impose severe risks and costs. Included are character assassinations—accusing those who speak out as "climate change deniers" who don't care about the environment; false branding as shills for Big Oil, tobacco companies or other non-existent sponsors; severe professional career penalties including unwarranted exclusions of significant research findings by scientific journals; lost research funding and promotion opportunities; employment terminations and sometimes even threats upon their personal safety and lives.

Dr. Singer, one of the world's most distinguished astrophysicists, is an exemplary hero to those of us who honor his unwavering commitment and enormous contributions to advance honest and informed science. In doing so, he has often fearlessly and effectively challenged the orthodoxy, just as open inquiry and adherence to sound scientific methods demand. Yes, and he has paid the price for such audacity, responding to disparaging attacks with quiet dignity and well-reasoned factual rebuttals.

Fred Singer generously honors me as well; this is made evident through his unfailing willingness to cross-check my article facts and frequently challenge my punctuation decisions. Here again he is also unfailingly correct in both regards.

Dear friend, thank you.

Preface

As WITH MY previous book, *Climate of Corruption: Politics and Power Behind the Global Warming Hoax,* I never planned to write this one either. The problem is that I'm a sucker for flattery, particularly when it comes from people I greatly admire. My very dear and distinguished friend Dr. Fred Singer—to whom I dedicate this book—often urged me to publish a compilation of material drawn from hundreds of articles I posted in my *Forbes* and *Newsmax* columns. Dr. Jay Lehr, another great and respected friend, has repeatedly done the same.

Like hemorrhoids, some itches are often better left unscratched, so I resisted the idea for quite some time. There were several reasons. One related to a time commitment/benefit factor. Posting an electronic blog column article involves days of effort and reaches hugely more readers than preparing a hard copy book which consumes months or even years with typically a far less expansive audience. Besides, people can read the stuff for free, which is just fine with me. Kindle is now rapidly changing this equation, and I think this is a very good thing.

Another big factor has to do with a major difference between much of my op-ed writing which most often addresses current national and world event-driven developments and a book which is directed to topics and perspectives with a much longer relevant "shelf life". This is particularly true regarding dynamic issues connected with environmental science and energy policy. After all, just as "they" say, climate really does change. Politics and alarmism on the other hand...not so much.

Also, I really didn't wish to write a book that was either just rehashed material in the earlier one or primarily amounting to a patchwork of old articles. And while quite a lot of overlap is unavoidable, my priority is to make the end product worth the reader's time investment. I hope you ultimately determine that it succeeds. The approach is an attempt to render a seamless narrative covering diverse, yet highly interconnected subjects, and also provide previously released articles in the Appendix section which provide additional elaboration.

I would like to point out—as I have before—that I have absolutely no financial affiliation with any interest group that might lend suspicion to my decidedly biased perspectives on topics I cover. Nope...no Big Oil or little oil connections...no research grants from special-interest sponsors...nadda. Nor have I ever claimed to be a "climate scientist", an ambiguous and increasingly compromised term in any case. Why and how this is occurring is a central theme of this book.

This should not be interpreted to suggest I don't hold objective and competent scientists in very high esteem. Quite the opposite, like other "space guys and gals", I deeply appreciate the importance of applying sound scientific methods to complex problems.

Yet, as Apollo 7 astronaut Walter Cunningham observed on the back cover of my *Climate of Corruption* book:

> *Those of us fortunate enough to have traveled in space bet our lives on the competence, dedication, and integrity of the science and technology professionals who made our missions possible...In the last twenty years, I have watched the high standards of science being violated by a few influential climate scientists, including some at*

NASA, while special interest opportunists have dangerously abused our trust.

Many former space program professionals are presently redirecting their expertise in complex systems design, model simulation and analyses to gain better understanding of natural climate dynamics. One independent group comprised primarily of retired NASA Apollo program veterans publishes non-feverish research conclusions on its website, TheRightClimateStuff.com.

Upon reviewing available data and scientific reports, the TRC concludes that many natural factors affect temperature variations of the Earth's surface and atmosphere, that human influences do not constitute "settled science", and that there is no reason for alarm about catastrophic warming based upon the outputs of non-validated computer models.

My Forbes article titled *A Cool-Headed Climate Conversation with Aerospace Legend Burt Rutan* reveals informed dismay regarding the sorry state of mainstream climate science by another person who has followed available data.[1]

Burt knows a thing or two about reliable research, having designed Voyager, the first aircraft to fly around the globe without refueling, SpaceShipOne which won the $10 million Ansari X-Prize for becoming the first privately-funded manned craft to enter the realm of space twice within a two-week period, and the Virgin GlobalFlyer which broke Voyager's time for a non-stop solo flight around the world.

Burt was shocked that to find there were climate scientists who refused to share raw data used to substantiate questionable and alarmist theories. He commented:

This was frankly astonishing because analyzing data is

---

[1] http://www.forbes.com/sites/larrybell/2012/09/09/a-cool-headed-climate-conversation-with-aerospace-legend-burt-rutan/

something I'm very good at. All my professional life I have been analyzing complex flight test data, interpreting it and presenting it. Something that I always did in flight test is to make a chart that shows every bit of the data, and only then, decide later on the basis of real observed results which parts of the data were valid.

Burt admonishes:

Tragically, policymakers have thrown horrendous amounts of taxpayer money needed for other purposes at solving an unsubstantiated emergency. It is scandalous that so many climate scientists who fully knew that Al Gore had no basis for his irresponsible claims stood mute. Meanwhile, that alarmism has generated billions of dollars more to finance a rapidly growing climate science industry with budgets that have risen by a factor of 40 since the early 1990s. I consider this failure to speak up just as unethical as the behavior of those who put out the false catastrophic claims.

My own professional skepticism regarding alarmist global warming claims was inadvertently introduced by Fred Singer, a space guy with true climate science credentials. While visiting my office a few years ago to exchange ideas about an entirely different subject, he happened to mention that satellite temperature recordings of the Earth's lower atmosphere were cooling more rapidly, relative to the surface, than CO_2 greenhouse theory predicts. It would be expected that carbon dioxide would warm the lower atmosphere first, which would then radiate heat back to the surface, the reverse of what was being observed.

I certainly had no reason to doubt him. Fred is an internationally recognized climate physicist and former Distinguished Research Professor at George Mason University. He

Larry Bell

served as the first director of the US Weather Satellite Service and also as vice chairman of the US National Advisory Committee on Oceans and Atmosphere. In addition, he founded the Science and Environmental Policy Project (SEPP) along with the Nongovernmental International Panel on Climate Change (NIPCC), another non-profit research group. Dr. Singer is the author, coauthor and editor of many books including *Climate Change Reconsidered* (several volumes), offering a comprehensive critique of the assessment reports of the United Nation's Intergovernmental Panel on Climate Change (IPCC) where he has served as an expert reviewer.

Fred's reference to fundamental disconnects between satellite records and what many of us were hearing about dire human influences upon climate piqued curiosity somewhat later. Although I had never previously been particularly interested in climate, media buzz about Al Gore's alarming *An Inconvenient Truth* movie and book was largely responsible for prompting me to direct considerable attention to the matter. That ongoing investigation grew out of my natural curiosity and space background which emphasizes holistic perspectives regarding basic principles that govern how natural and technical systems are connected, and how they can be managed to support the most complex systems of all—we humans.

The more I delved into the purported science and facts behind the alarmism, the more convinced I became that the real inconvenient truth pointed to massive deceptions. Some of those became evident through releases of scandalous ClimateGate emails that revealed corrupt collusions among leading IPCC scientists to drive ideological and political agendas. Another big light bulb flashed in my awareness upon reading statements presented by a very wealthy green energy hedge fund investor and lobbyer poised to make windfall profits selling CO_2 offsets if and when cap-and-trade was passed. Speaking before a 2007 Joint House Hearing of the Energy Science Committee, climate doom prophet Al Gore told members:

> *As soon as carbon has a price, you're going to see a wave [of investment] in it...There will be unchained investment.*

Those unwelcome and startling insights sparked a continuing priority to identify and expose agenda-driven fear-mongering masquerading as science. Briefly summarized, but elaborated in later pages, my key conclusions follow:

- No sane person will deny that dramatic, often abrupt climate changes have occurred throughout our planet's history.

- No informed person will believe climate change events are more frequent or severe since the Industrial Revolution...the past few decades in particular.

- No honest person will claim confidence in an ability to predict the long-term future, warmer or cooler, for better or worse, based upon highly theoretical climate models which entirely omit poorly understood influences.

- No honorable scientist will compromise objective research principles through omissions of unknowns and uncertainties in order to influence political policy agendas.

- And no concerned citizen should be willing to tolerate abuses of science and public trust. Climate alarmists support unwarranted economic burdens which fall heaviest upon those least able to afford them.

Section 1

The Climate Alarm Industry

ROYAL SOCIETY OF Chemistry Fellow Dr. Leslie Woodcock observes that green lobbies use unwarranted climate alarm to support a very costly "do-good" industry. He recently told Britain's *Yorkshire Evening Post*:

> *If you talk to real scientists who have no political interests, they will tell you there is nothing in global warming. It's an industry which creates vast amounts of money for some people.*

Yes, lots of money.

Those many billions fund the growth of government regulatory agencies that depend upon public fear; university departments that bend objectivity to secure research grants; activist environmental groups that rely upon crisis-premised donations to support lobbying and media programs; anti-fossil "alternative energy" lobbies seeking special subsidies; and a wide host of politicians, prophets and profiteers who cash in on "save the world" hype to fill campaign coffers and personal bank accounts.

Perhaps unsurprisingly, Al Gore immediately comes to mind as an example. Consider that it was only about a dozen years after three decades of global cooling when some prominent scientists were predicting an arrival of the next Ice Age that then-Senator Gore convened his famous 1988 Senate Committee on Science, Technology and Space hearings which produced a man-made global warming crisis media frenzy.

As his colleague Senator Timothy Wirth who helped organize the meetings later stated in a *PBS* interview:

> *We called the Weather Bureau and found out what historically was the hottest day of the summer...so we scheduled the hearing that day, and bingo, it was the hottest day on record in Washington, or close to it...we went in the night before and opened all the windows so that the air conditioning wasn't working inside the room.*

That very same Al Gore began to amass a large fortune through holdings in companies which were "going green". By 2008 he was able to put together $35 million into hedge funds and private partnerships through the Capricorn Investment Group founded by his Canadian billionaire buddy Jeffrey Skoll, the first president of EBay Inc.

Climate crusader Gore was also poised to make windfall profits selling CO_2 offsets through his stake in the Chicago Climate Exchange if and when Congress passed cap-and-trade legislation he promoted. A 2010 Republican Housecleaning swept away those cap-and-trade legislation hopes.

And whereas an earlier presidential candidate Gore had run on a green platform which included ethanol tax breaks, he subsequently admitted that maybe this wasn't entirely about saving the planet after all. Speaking in 2010 at a green energy business conference in Athens, Greece, he said, "It is not a good policy to have these massive subsidies for first-generation ethanol."

He then explained to *Reuters*:

> *One of the reasons I made that mistake is that I paid particular attention to the farmers in my home state of Tennessee, and I had a certain fondness for the farmers in the state of Iowa [the first-in-the-nation caucuses state] because I was about to run for president.*

Nobel Laureate physicist Dr. Ivar Giaever has referred to global warming ideology as a "pseudoscience" that begins with an emotionally-appealing hypothesis, and "then only looks for items which appear to support it," while ignoring ample contrary evidence.

Tragically, that pseudoscience does the greatest injustice to those who can least afford it. These penalties come in forms of green energy subsidies, domestic fossil development impediments, and runaway EPA regulatory policies which drive up fuel and electricity costs, food prices, federal debt, and monetary inflation.

And why are these scams so successful? Leslie Woodcock explains that:

> *...you can't blame people with no science education for wanting to be seen to be good citizens who care about their grandchildren's future and the environment.*

Climate science industrialists and eco-elitist zealots arguing that fossil-fueled economic growth is the enemy of the environment miss a vital point. They overlook the fact that that such progress yields technological innovation and prosperity essential to support more resourceful, cleaner and healthier lifestyles.

Meanwhile, as global mean temperatures have remained flat now for going on two decades despite increased atmospheric CO_2

levels, the UN's Intergovernmental Panel on Climate Change (IPCC) and the US. EPA continue to pitch this tiny trace greenhouse gas as a climate-ravaging menace and ignore enormous benefits this essential plant food affords.

Chapter One: That Scientific Climate Crisis Consensus...Not!

IF I WERE to ask if you believed in "climate change" or "global warming", how would you interpret the question? Would you assume I wondered if you thought the climate was really becoming dangerously overheated and that we humans, via fossil fuel burning in particular, are causing it? Would you suppose I wanted to know if you imagined the world has been warming over the past century or more, and maybe a range of human activities are having some influence...and possibly, even for the better? Or, maybe, might I be questioning if you knew that the Earth has gone through dramatic warming and cooling changes over millions of years before we humans arrived on the scene...agriculture, smokestacks, SUV's and all?

Perhaps you recognize that each of these personal interpretations of the question might evoke a "yes" answer. If so, for starters, please raise your hand if you don't believe—as rumored—that climate changes. What? No one? Great! That's a big relief. I don't see any of those climate change "deniers" we hear so much about among us. And now that we seem to have a consensus so far that most all of us "believe in global warming" (and cooling), AKA "climate change", let's break this issue down a bit more regarding exactly what we agree about.

Chapter One: That Scientific Climate Crisis Consensus

What about whether or not we believe humans have any influence on climate? I'm talking zero here, none, zip, nadda. Can anyone be sure we don't have absolutely any, even though that amount has never, not ever, been measured?

No one has ever been able to measure human contributions to climate. Don't even think about buying a used car from anyone who claims they can.

Still, wouldn't you imagine that just about everything we do, including land use changes such as agriculture, would have some, albeit teensy weensy, influence at least upon local, seasonal weather...which, in turn, is part of a humongous global multi-decadal climate picture? I'm not asking whether that influence, however miniscule, is for better or worse. That's another ball of wax altogether. So can't we agree that we mortals have some influence on warming (and probably cooling) too?

Are we still together so far?

For example, consider those carbon dioxide emissions that we keep hearing about. Who can doubt that CO_2 is a "greenhouse gas" that absorbs energy in the thermal infrared portion of the electromagnetic spectrum whereby more energy is transferred from the surface to the atmosphere through latent heat long-wave radiation?

Yet even the UN's IPCC which claims to be the ultimate authority on all things climate admits that its past estimates of climate sensitivity to radiatively-active gases were exaggerated and that important effects of cloud cover and natural ocean cycle changes are unknown.

Now for the biggie...just how worried should we be? Are WE causing a climate catastrophe?

If you believe so, then based on what evidence? Climate models? Have they ever predicted anything right so far...including flat mean global temperatures at a time that atmospheric CO_2 concentration is at a much-touted record high for recent times?

A Timely Climate Change Perspective

Yes, climate definitely changes. Cyclical, abrupt and dramatic global and regional temperature fluctuations have occurred over millions of years, long before humans invented agriculture, industries, internal combustion engines and carbon-trading schemes. Many natural factors are known to contribute to these changes, although even the most sophisticated climate models and theories they are based on cannot predict the timing, scale (either up or down), or future impacts—much less the marginal contributions of various human influences.

And while global warming has been trumpeted as an epic climate change crisis with human-produced CO_2, a trace atmospheric "greenhouse gas" branded as a primary culprit and endangering "pollutant", don't be too sure about the veracity of those pitches. Throughout earlier periods of Earth's history those levels have been many times higher than today, with temperature changes proceeding, not following atmospheric CO_2 changes. It doesn't require a degree in a climate science, or rocket science either for that matter, to understand these basic facts.[2]

Fossil records reveal that atmospheric CO_2 levels around 600 million years ago were about 7,000 parts per million, compared with 400ppm now. Then approximately 480 million years ago those levels gradually dropped to 4,000ppm over about 100 million years, while average temperatures remained at a steady 72 degrees Fahrenheit. They then jumped rapidly to 4,500ppm and guess what! Temperatures dove to an estimated average similar to today, even though the CO_2 level was around twelve times higher than now. Yes, as CO_2 went up, temperatures plummeted.

About 438 million years ago, atmospheric CO_2 dropped from 4,500ppm to 3,000ppm, yet according to fossil records,

---

[2] Kevin Trenberth, *Predictions of Climate*, Climate Feedback: The Climate Change Blog, June 2007

world temperatures shot rapidly back up to an average 72°F. So regardless whether CO_2 levels were 7,000ppm or 3,000ppm, temperatures rose and fell independently.[3]

Over those past 600 million years there have been only three periods, including now, when Earth's average temperature has been as low as 54°F. One occurred about 315 million years ago during a 45-million-year-long cool spell called the Late Carboniferous period which established the beginning of most of our planet's coalfields. Both CO_2 and temperatures shot back up at the end of it just when the main Mesozoic dinosaur era was commencing. CO_2 levels rose to between 1,200ppm and 1,800ppm, and temperatures again returned to the average 72°F that Earth seemed to prefer.[4]

Around 180 million years ago, CO_2 rocketed up from about 1,200ppm to 2,500ppm. And would you believe it? This coincided again with another big temperature dive from 72°F to about 61°F. Then at the border between the Jurassic period when T. Rex ruled and the Cretaceous period that followed, CO_2 levels dropped again, while temperatures soared back to 72°F and remained at that level (about 20 degrees higher than now) until long after prodigious populations of dinosaurs became extinct. And flatulent as those creatures may possibly have been, at least there is no evidence that they burned coal or drove SUVs.

Based upon a variety of proxy indicators, such as ice core and ocean sediment samples, our planet has endured large climate swings on a number of occasions over the past 1.5 million years due to a number of natural causes. Included are seasonal warming and cooling effects of plant growth cycles, greenhouse gases and

---

[3] Robert Berner and Zawert Kothavala, *GeoCarb III: A revised Model of Atmospheric CO₂ over Phanerozoic Time*, American Journal of Science, vol. 301 (February 2001)

[4] Ian Wishart, AirCon, *The Seriously Inconvenient Truth about Global Warming*, HATM Publishing, 2009, 35

aerosols emitted from volcanic eruptions, Earth orbit and solar changes, and other contributors with combined influences. Yet atmospheric CO_2 levels have remained relatively low over the past 650,000 years—even during the six previous interglacial periods when global temperatures were as much as 9°F warmer than temperatures we currently enjoy.[5][6]

Over the past 400,000 years, much of the Northern Hemisphere has been covered by ice up to miles thick at regular intervals lasting about 100,000 years each. Much shorter interglacial cycles like our current one lasting 12,000 to 18,000 years have offered reprieves from bitter cold. Yes, from this perspective, current temperatures are abnormally warm. By about 12,000 to 15,000 years ago Earth had warmed enough to halt the advance of glaciers and cause sea levels to rise, and the average temperature has gradually increased on a fairly constant basis ever since, with brief intermissions.[7]

During a period from about 750BC to 200BC, before the founding of Rome, temperatures dropped and European glaciers advanced. Then the climate warmed again, and by 150BC grapes and olives were first recorded to be cultivated in northern Italy. As recently as 1,000 years ago (during the "Medieval Warm Period"), Icelandic Vikings were raising cattle, sheep and goats in grasslands on Greenland's southwestern coast.

Then, around 1200, temperatures began to drop, and Norse settlements were abandoned by about 1350. Atlantic pack ice began to grow around 1250, and shortened growing seasons and unreliable weather patterns, including torrential rains in Northern

---

[5] *MIT Researcher Finds Evidence of Ancient Climate Swings*, Science Daily, April 20, 1998

[6] Ernst-George Beck, *180 Years of Atmospheric CO_2 Gas Analysis by Chemical Methods*, Energy and Environment, vol. 18, no, 2 2007

[7] Roy W. Spencer, *Climate Confusion*, Encounter Books 2008, 87

Europe led to the "Great Famine" of 1315-1317.[8][9]

Temperatures dropped dramatically in the middle of the 16th century, and although there were notable year-to-year fluctuations, the coldest regime since the last Ice Age (a period termed the "Little Ice Age") dominated the next hundred and fifty years or more. Food shortages killed millions in Europe between 1690 and 1700, followed by more famines in 1725 and 1816. The end of this time witnessed brutal winter temperatures suffered by Washington's troops at Valley Forge in 1777, and Napoleon's bitterly cold retreat from Russia in 1812.[10][11]

Although temperatures have been generally mild over the past 500 years, we should remember that significant fluctuations are normal. The past century has witnessed two distinct periods of warming. The first occurred between 1900 and 1945, and the second, following a slight cool-down began quite abruptly in 1975. That second period rose at quite a constant rate until 1998, and then stopped and began falling again after reaching a high of 1.16°F above the average global mean.

See Appendix 1-1: Some scientists dispute that there has been any warming since 1978. *Any Global Warming Since 1978? Two Climate Experts Debate This*, Larry Bell, Forbes Opinions, June 18, 2013.

About half of all estimated warming since 1900 occurred before the mid-1940s despite continuously rising CO_2 levels. Even key ClimateGate figure UK East Anglia University Climatic Research Center Director Phil Jones admitted back in 2010 that

---

[8] S. Fred Singer and Dennis T. Avery, *Unstoppable Global Warming*, Rowman &Littlefield, 2007, 138

[9] U.S. Environmental Protection Agency, *Global Warming-Climate*, October 14, 2004

[10] Syun-Ichi Akasofu, *Two Natural Components of Recent Climate Change*, March 30, 2009

[11] Singer and Avery, Ibid

there had been no statistically significant warming for at least 15 years. He has also admitted that temperatures during the Middle Ages may have been higher than today.[12][13][14]

So perhaps you'll wish to ponder this question: Given that over most of the Earth's known climate history the atmospheric CO_2 levels have been between four and eighteen times higher than now—throughout many times when life not only survived but also flourished; times that preceded humans; times when CO_2 levels and temperatures moved in different directions…how much difference will putting caps on emissions accomplish? Consider also that about 97% of all current atmospheric CO_2 derives from natural sources.

And yes again, change is the true nature of climate. After all, if climate didn't change, we really wouldn't need a word for it would we? Wouldn't it all just be weather?

But What About that "Settled" Climate Debate?

Remember hearing that 97% of all scientists agree that fossil-fueled smokestacks and SUVs are melting the glaciers, raising the oceans and generally tilting the planet at a calamitous tipping point? Gosh, if climate scientists can't be trusted to conduct a reliable opinion poll of their own members, what hope is there for those of us who are mere rocket scientists?

Any alarmist claims suggesting most all scientists agree we humans are setting the world on fire are hot-headed nonsense.

---

[12] Graeme Stephens, *Cloud Feedbacks in the Climate System: A Critical Review*, Journal of Climate. Vol. 18, January 15, 2005, 237-73
[13] *Peer Reviewed Study Rocks Climate Debate! Nature Not Man Responsible for Recent Global Warming*, Climate Depot, July 22, 2009
[14] Roy Spencer, John Christy, et al., *Cirrus Disappearance: Warming Might Thin Heat-Trapping Clouds*, University of Alabama-Huntsville, August 9, 2007

That endlessly trumpeted baloney can be traced to two primary sources.

The earliest originated from a 2009 American Geophysical Union (AGU) survey consisting of an intentionally brief two-minute, two question online survey sent to 10,257 Earth scientists by two researchers at the University of Illinois-Chicago.

That anything-but-scientific survey asked two questions. The first:

> *When compared with pre-1800s levels, do you think that mean global temperatures have generally risen, fallen, or remained relatively constant?*

Few would be expected to dispute this since the planet began thawing out of that Little Ice Age in the middle 19th century which predated the Industrial Revolution. (That was the coldest period since the last real Ice Age ended roughly 10,000 years ago.)

The second question asked:

> *Do you think human activity is a significant contributing factor in changing mean global temperatures?*

So what constitutes "significant"? Not questioned was whether "changing" included both cooling and warming...or whether for "better" and "worse"?

And which contributions? They also weren't asked whether they believed the anthropogenic (human-caused) contribution to global warming was or might become sufficient to warrant concern, or the adoption of government regulatory policies.

Of the 3,146 who responded (a 31% return rate), only a small subset of just 77 (2.5%) were represented in the survey statistic. These are ones who listed "climate science" as their area of expertise and had been successful in getting more than half of their papers recently accepted by peer-reviewed climate science journals. In other words, that "97% all scientists" referred to a laughably puny number of 75 of those 77 who answered "yes".

Get that…of the 3,146 Earth scientists who responded, 97% of the cherry-picked 2.5% who were counted in the survey agreed that humans have at least some influence on climate! That's really a ton of consensus!

That unsupportable 97% climate scientist consensus pitch was concocted again from a study produced by a Global Change Institute researcher which supposedly looked at about 12,000 scientific climate-related abstracts published from 1991-2011.

Annually tracked survey tabulations revealed that the percentage of respondents indicating "no opinion" grew from more than 50% to more than 60% over the study period. Meanwhile, the abstracts that expressly supported the anthropogenic global warming (AGW) theory fell from around 50% to well below 40%. Only 65 of those 12,000 papers showed a strong AGW endorsement.

It's quite a stretch to conclude that an "overwhelming percentage" (allegedly 97.1% based upon abstract ratings) supported the AGW theory, given that less than 40% of the abstracts explicitly did so.

A much more credible 2012 survey by the American Meteorological Society found that only one in four respondents agreed with a UN-IPCC assertion that humans are primarily responsible for recent warming. And while 89% believed that global warming is occurring, only 30 percent said they were worried.

Chapter Two: Hot and Cold Sweats

CLIMATE ALARMISM HAS run hot and cold for a long time, although not necessarily beginning in that sequence. Like for example during a period from about 1940 to the early 1970s when records showed a cooling trend and many scientists predicted that the Earth was heading for the next in a regular series of Ice Ages. The popular press, including *Time*, *Newsweek*, and the *New York Times*, featured these claims in a number of alarming headline articles.

Remarkably, within only about half of a commonly accepted 30-year-long climate period later media attention shifted to a new and opposite threat...the one that set Al Gore's pants on fire during his 1988 Senate hearings on the matter. By that time the United Nations had already determined that global warming was a crisis and that human fossil fuel CO_2 emissions were the cause. In short order they established a cap-and-trade program (the Kyoto Protocol) to tax carbon emissions, plus demanded additional economic penance from developed countries for all that climate damage their unfair prosperity is causing.

Then everything ran into an unanticipated snag...that "best laid plans of mice and men going awry" conundrum thing. While the IPCC's climate models were predicting carbon dioxide-driven temperatures going orbital and sea levels flooding Capitol Hill, something went terribly wrong. Yup, the real temperature trajectory went flat, and has remained that way since the time before most of today's high school students were born.[15]

Anyone who says they can confidently predict global climate changes or effects is either a fool or a fraud. No one can even forecast global, national or regional weather conditions that will occur months or years into the future, much less climate shifts that will be realized over decadal, centennial and longer periods. Nevertheless, this broadly recognized limitation has not dissuaded doomsday prognostications that have prompted incalculably costly global energy and environmental policies. Such postulations attach great credence to computer models and speculative interpretations that have no demonstrated accuracy.

The primary source of scary climate predictions endlessly bleated in the media originates from politically cherry-picked summary reports issued by the UN's IPCC. Yet even its 2001 report chapter titled *Model Evaluation* contains this confession:

> *We fully recognize that many of the evaluation statements we make contain a degree of subjective scientific perception and may contain much 'community' or 'personal' knowledge. For example, the very choice of model variables and model processes that are investigated are often based upon subjective judgment and experience of the modeling community.*[16]

In that same report the IPCC further admits:

> *In climate research and modeling, we should realize that we are dealing with a coupled non-linear chaotic system, and therefore that the long-term prediction of future*

---

[15] *Climate Forecasting Models: Not Pretty—Not Smart*, Larry Bell, Forbes Opinions, August 9, 2011
[16] IPCC Climate Change 2001, Chapter 8. *Model Evaluation*, 475

climate states is not possible.

Here, while openly acknowledging that their models can't be trusted, they obviously need to apply them to justify their budgets and influence. Without contrived, frightening forecasts the IPCC would soon be out of business.

So Let's Take a Closer Look at those Models

The 2007 report changed IPCC's story significantly, placing "great confidence" in the ability of General Circulation Models (GCMs) to responsibly attribute observed climate change to anthropogenic greenhouse gas emissions. It stated that:

> ...*climate models are based on well-established physical principles and have been demonstrated to reproduce observed features of recent climate...and past changes.*[17]
> [18]

Yet even Dr. Kevin Trenberth, a lead author of 2001 and 2007 IPCC report chapters, has admitted that the IPCC models have failed to duplicate realities. Writing in a 2007 *Predictions of Climate* blog appearing in the science journal *Nature.com* he stated:

> *None of the models used by the IPCC are initialized to the observed state and none of the climate states in the models correspond even remotely to the current observed state.*[19]

---

[17] *Climate Change Reconsidered; The Report of the Nongovernmental Panel on Climate Change*, 2009, p 9.

[18] IPCC, 2007-I, p 591

[19] Trenberth quote, Kevin Trenberth, *Predictions of Climate*, Climate Feedback: The Climate Change Blog, June 2007

Dr. Syun-Ichi Akasofu, the former director of the International Arctic Research Center at the University of Alaska-Fairbanks, has determined that IPCC computer models have not even been able to duplicate observed temperatures in Arctic regions. While the atmospheric CO_2 forecasts indicated warm Arctic conditions, they were lower than actually reported, and colder areas were absent. Dr. Akasofu stated:

> *If fourteen GCMs [General Circulation Models] cannot reproduce prominent warming in the continental Arctic, perhaps much of this warming is not produced by greenhouse effect at all.*[20]

Dr. Graeme Stephens at the Colorado State University's Department of Atmospheric Science warned in a 2008 paper published in the *Journal of Climate*, that computer models involve simplistic cloud feedback descriptions:

> *Much more detail on the system and its assumptions [is] needed to judge the value of any study. Thus, we are led to conclude that the diagnostic tools currently in use by the climate community to study feedback, at least as implemented, are problematic and immature and generally cannot be verified using observations.*[21]

Prominent scientist, the late Dr. Joanne Simpson, developed some of the first mathematical models of clouds in attempts to better understand how hurricanes draw power from warm seas. Ranked as one of the world's top meteorologists, she believed that

---

[20] Syun-Ichi Akasofu, *Two Natural Components of the Recent Climate Change*, March 30, 2009

[21] *Graeme Stephens, Cloud Feedbacks in the Climate System: a Critical Review*, Journal of Climate, vol. 18, January 15, 2005, 237-73

global warming theorists place entirely too much emphasis upon faulty climate models, observing:

> *We all know the frailty of models concerning the air-surface system...We only need to watch the weather forecasts [to prove this].*[22]

Dr. Roy Spencer, a principal research scientist at the University of Alabama-Huntsville and former senior scientist for climate studies at NASA, has observed that results of the one or two dozen climate modeling groups around the world often reflect a common bias. One reason is that many of these modeling programs are based upon the same "parameterization" assumptions; consequently, common errors are likely to be systematic, often missing important processes.

Such problems arise because basic components and dynamics of the climate system aren't understood well enough on either theoretical or observational grounds to even put into the models. Instead, the models focus upon those factors and relationships that are most familiar, ignoring others altogether. As Dr. Spencer notes:

> *Scientists don't like to talk about that because we can't study things we don't know about.*[23]

A peer-reviewed climate study that appeared in the July 23, 2009 edition of *Geophysical Research Letters* went even farther in its characterization of faulty climate modeling practices. The paper noted IPCC modeling tendencies to fudge climate projections by exaggerating CO_2 influences and underestimating the importance

---

[22] *Pioneer Meteorologist Unearthed Mysteries of Clouds*, Storms, Wall Street Journal, March 2, 2010.
[23] Roy W. Spencer, *Climate Confusion*, Encounter Books, 2008, 87

of shifts in ocean conditions.

The research indicated that influences in solar changes and intermittent volcanic activity have accounted for at least 80% of observed climate variation over the past half century. Study coauthor Dr. John McLean observed:

> *When climate models failed to retrospectively produce the temperatures since 1950, the modelers added some estimated influences of carbon dioxide to make up the shortfall.*

He also highlighted inabilities of computer models to predict El Niño ocean events which can periodically dominate regional climate conditions, hence further reducing model meaningfulness.[24]

Dr. J. Scott Armstrong, a professor at the Wharton School, University of Pennsylvania and leading expert in the field of professional forecasting, believes that prediction attempts are virtually doomed when scientists don't understand or follow basic forecasting rules. He and colleague Dr. Kesten Green of Monash University conducted a "forecasting audit" of the 2007 IPCC report and "found no references...to the primary sources of information on forecasting methods" and that "the forecasting procedures that were described [in sufficient detail to be evaluated] violated 72 principles. Many of the violations were, by themselves, critical".

A fundamental principle that IPCC violated was to "make sure forecasts are independent of politics". Armstrong and Green observed that:

> *The IPCC process is directed by non-scientists who have*

---

[24] *Peer-Reviewed Study Rocks Climate Debate! Nature Not Man Responsible for Recent Global Warming*, Climate Depot, (July 22, 2009).

policy objectives and who believe that anthropogenic global warming is real and a danger.

They concluded that:

The forecasts in the report were not the outcome of scientific procedures. In effect, they were the opinions of scientists transformed by mathematics and obscured by complex writing...We have not been able to identify any scientific forecasts of global warming. Claims that the Earth will get warmer have no more credence than saying it will get colder.

Kevin Trenberth argued in his 2007 *Nature* blog that "the IPCC does not make forecasts", but "instead proffers 'what if' projections that correspond to certain emission scenarios"; and then hopes these "projections...will guide policy and decision makers." He went on to say, "there are no such predictions [in the IPCC reports] although the projections given by the Intergovernmental Panel on Climate Change (IPCC) are often treated as such. The distinction is important."

Armstrong and Green challenge that semantic defense, pointing out that "the word 'forecast' and its derivatives occurred 37 times, and 'predict' and its derivatives occurred 90 times in the body of Chapter 8" of [the IPCC's 2007] the Working Group I report".[25][26]

Of course there would be very little interest in model forecasts at all if it were not for hysterical hype about a purported man-made climate crisis caused by carbon dioxide fossil fuel

---

[25] *Climate Change Reconsidered; The Report of the Nongovernmental Panel on Climate Change*, 2009, p 10-1.

[26] Green, K.C. and Armstrong, J.S. 2007. *Global warming forecasts by scientists versus scientific forecasts.* Energy Environ. 18: 997-1021

emissions. Without CO_2 greenhouse gas demonization there is no basis for cap-and-tax schemes, unwarranted "green" fuel subsidies, expansion of government regulatory authority over energy production and construction industries through unintended misapplications of the Clean Air Act, claims of polar bear endangerment to prevent drilling in ANWR, or justifications for massive climate research budgets including...guess what? Yup! Lots of money to produce more climate model forecasts that perpetuate these agendas.

One of the exposed ClimateGate emails reveals Dr. Peter Thorne of the UK Met Office wisely observing:

> *I also think the science is being manipulated to put a political spin on it which for all our sakes might not be too clever in the long run.*

Another scientist worries:

> *...clearly, some tuning or very good luck [is] involved. I doubt the modeling world will be able to get away with this much longer.*

Of course any model can only be as good as the information that is programmed into it. As Ian "Harry" Harris reports:

> *[The] hopeless state of their [CRU] database. No uniform data integrity. It's just a catalogue of issues that continues to grow as they're found...There are hundreds if not thousands of pairs of dummy [surface temperature recording] stations...and duplicates...Aarrggghh! There truly is no end in sight. This project is such a MESS. No wonder I needed therapy!!* [27]

---

[27] Energy Tribune, *Is It Really The Warmest Ever?*, Joseph D'Aleo, January

Still another modeler observed:

> *It is inconceivable that policymakers will be willing to make billion-and trillion-dollar decisions for adaptation to the projected regional climate change based on models that do not even describe and simulate the processes that are the building blocks of climate variability.*[28]

Yes...and that's the biggest climate crisis of all.

29, 2011

[28] Sources (Email Quotations): *Climategate 2: The Scandal Continues*, Myron Ebell, CEI Press Release

ChapterThree: ATree Ring Circus

IF GLOBAL WARMING isn't the *Greatest Show on Earth*, it's certainly the costliest and most bizarre. An early act featuring a hockey stick-shaped graph published by IPCC in 2001 profoundly influenced world energy and environmental policies.

Based heavily upon data taken from tree growth rings on the Yamal Peninsula in Siberia, it indicated that world temperatures which had been stable for 900 years until the 20th century suddenly soared due to human fossil fuel-burning greenhouse gas emissions—at least that was the IPCC's story.

Fallen Hot Aerialist Returns to Center Ring

Although science behind that hockey stick chart has now been thoroughly challenged, its creator, Dr. Michael Mann, is a harsh critic of skeptics who dare to question the existence of the crisis he has failed to prove. His January 15 *New York Times* Op/Ed column titled *If You See Something, Say Something* charges that despite an overwhelming consensus among climate scientists that human-caused climate change is happening, a "virulent strain of anti-science infects the halls of Congress, the pages of a few leading newspapers and what we see on TV, leading to the appearance of a debate where none should exist."[29]

Having already concluded not only that global warming is dangerous, that human emissions pose that threat, Mann then urges "mainstream scientists" (purportedly all of those who agree with him) to get directly involved in remedial technology and policy activism. Such involvement includes determining whether to go "full-bore" on nuclear power, whether to invest in and deploy renewable wind, solar and geothermal energy on a huge scale, and whether to price carbon emissions through cap-and-trade legislation or by imposing a carbon tax.

Mann refers to the late Stanford University Professor Stephen Schneider, a fellow man-made global warming advocate, as a good example. Incidentally, this is the same Stephen Schneider who authored *The Genesis Strategy*, a 1976 book warning that global cooling risks posed a threat to humanity. Schneider later changed that view 180 degrees, serving as a lead author for important parts of three IPCC reports.

Blurring the divide between objective science and political science, Schneider once said:

> On the one hand, as scientists we are ethically bound to the scientific method. On the other hand, we are not just scientists, but human beings as well. And like most people, we'd like to see the world a better place, which in this context translates into our working to reduce the risk of potentially disastrous climatic change. To do that we need to get broad-based support, to capture the public's imagination. That, of course, entails getting loads of

---

[29] *If You See Something, Say Something*, Michael Mann, January 17, 2014, New York Times
http://www.nytimes.com/2014/01/19/opinion/sunday/if-you-see-something-say-something.html?_r=0

media coverage. So we have to offer up scary scenarios,
make simplified, dramatic statements, and make little
mention of any doubts we might have.

Canadians Put Hockey Stick on Ice

Scenarios offered up in Michael Mann's hockey stick chart clearly met Stephen Schneider's criteria. They were simple, scary and dramatic, scored loads of media coverage, and its advocates made no mention of doubts. Instead, the doubts which ultimately led to its ultimate infamy came from Steven McIntyre and Ross McKitrick, two outside Canadian statisticians.

Upon careful investigation, McIntyre and McKitrick discovered fundamental statistical method errors so profound that even random numbers fed into Mann's program would produce a hockey stick curve. That wasn't all. The Medieval Warm Period which occurred about one thousand years ago and the Little Ice Age between about 1300-1850 somehow turned up missing.

And as for those Yamal tree samples, they came from only 12 specimens of 252 in the data set...while a larger data set of 34 trees from the same vicinity that weren't used showed no dramatic recent warming, but warmer temperatures in those Middle Ages.

Scientific critics have also raised another looming question. Since Mann's 1,000-year-long graph was cobbled together using various proxy data derived from ice cores, tree rings and written records of growing season dates up until 1961 where it then switched to using surface (ground station) temperature data, then why change in 1961? Some theorize that maybe it's because that's when other tree ring proxy data calculations by Keith Briffa at the East Anglia University Climate Research Unit (CRU) began going the other way in a steady temperature decline.

After presenting these unwelcome results to Mann and others, Briffa was reportedly put under pressure to recalculate them. He did, and the decline became even greater. As recorded in ClimateGate emails, this presented what Mann referred to as a "conundrum" in that the late 20th century decline indicated by Briffa would be perceived by IPCC as "diluting the message", that there was a "problem", and that it posed a "potential distraction/detraction". Mann went on to say that the warming skeptics would have a "field day" if Briffa's declining temperature reconstruction was shown, and that he would "hate to be the one" to give them "fodder".

In an email sent to Mann and others, CRU's director Dr. Philip Jones reported:

> I've just completed Mike's [Mann's] Nature [journal] trick of adding in the real temps to each series for the last 20 years (i.e., from 1981 onwards) and from 1961 for Keith's [Briffa's] to hide the decline [in global temperatures]...

Then all of the proxy and surface measurement chartings were presented in different colors on a single graph, and Briffa's were simply cut off in a spaghetti clutter of lines at the 1961 date.[30]

The ClimateGate Probe Dog and Pony Shows

Deluged with accusations of wrongdoing in the wake of ClimateGate email revelations, Mann's employer, Penn State University appointed an Inquiry Committee to investigate the matter. The committee was charged with looking into four types of allegations—whether he directly or indirectly:

1) suppressed or falsified data;
2) subsequently deleted or destroyed emails or other

---

[30] *Mike's Nature Trick*, Watts Up With That, November 20, 2009

information;

3) misused privileged or confidential information; and

4) engaged in any activities that "seriously" deviated from accepted academic practices.

According to the committee report, after interviewing Mann for "nearly two hours", he disclaimed all allegations. Members also "culled through" about 1,075 CRU emails to identify those sent by Mann, sent or copied to him, or which discussed him.

Of particular interest was a request from Phil Jones asking Mann to delete email records being sought under the UK's Freedom of Information Act and to get a colleague, Eugene Wahl, to do the same. Mann had then replied, "I'll contact Gene about this ASAP".

PSU investigators never chose to interview Wahl who later testified to a federal inspector general that he did receive Mann's message and complied with the deletions. Since there are no records to prove otherwise, everyone is asked to take Mann's word that he didn't do the same.[31]

In conclusion, the Committee found "no substance" in the first three allegations, and that there was "no basis for further examination" of those. For example, they determined that the "trick" referred to in Phil Jones's November 16, 1999 email was nothing more than "a statistical method used to bring together two or more different kinds of data sets together in a legitimate fashion by a technique that has been reviewed by a broad array of peers in the field".

One of the invited outside reviewers interviewed was astounded at the Committee's conclusions. When Dr. Richard Lindzen, a professor in the Department of Earth, Atmospheric and Planetary Sciences, Massachusetts Institute of Technology,

[31] *Michael Mann And The ClimateGate Whitewash: Part One*, Larry Bell, June 28, 2011, Forbes

was informed during the interview that the first three allegations had already been dismissed at the inquiry stage, his response, as quoted in the Committee's report, was:

> *It's thoroughly amazing. I mean these are issues that he explicitly stated in the emails. I'm wondering what is going on?*

In the final analysis, while Michael Mann was clearly *not* found guilty of wrongdoing, neither was the quality of his science or alarming research conclusions validated. It should be instructive to note, however, that the IPCC has now obviously distanced itself from his hockey stick chart which it so prominently featured in their headline-grabbing *2001 Assessment Report*.

See Appendix 1-2 for more discussion of the broken hockey stick matter: *Climate Gate Whitewash-Part I: Michael Mann Getting Heat in Grillings of PSU and UVA*, Larry Bell, Forbes Opinions, June 28, 2011

Locking the ClimateGate behind Them

But what about the IPCC and their key network of researchers within the UK's University of East Anglia (UEA)-Climate Research Unit (CRU)? Were they ever found innocent of any culpability in the ClimateGate scandal?

Three UK-based inquiries, each with transparent damage-control overtones, yielded little to support scientific confidence. Two were "independent" internal self-investigations that were launched by UEA. The third was a cursory, narrowly-focused inquiry conducted by the British House of Commons' Science and Technology Select Committee.[32]

---

[32] *Michael Mann And The ClimateGate Whitewash, Part II*, Larry Bell, July 5, 2011

The scientific misconduct charges against key IPCC and its CRU participants included failures to provide a full and fair view to policymakers and the IPCC of all available evidence; deliberately obstructing access to data and methods to those with opposing viewpoints; failures to comply with Freedom of Information Act (FOIA) requirements; and coordinated efforts to influence review panels of prestigious journals to block papers presenting rival scientific findings from being published.

Regarding the "Parliamentary Inquiry" undertaken by the House of Commons' Science and Technology Select Committee, an in-depth investigation was out of the question because of severely constrained time due to an upcoming election. Recognizing that it would not "be able to cover all of the issues raised by the events at UEA", questioning of witnesses was limited to a single day and the inquiry scope was limited to three key areas: freedom of information issues; accuracy and availability of CRU data and programs; and the independent reviews.[33]

Taking no direct testimony from those who challenged CRU activities, methods or errors, the committee nevertheless determined that there was essentially nothing wrong with the organization's basic science. Then mistakenly assuming that important investigations they had no time or expertise to conduct would be fully covered by the other "independent" reviews which never occurred, they simply endorsed IPCC's alarmist human-caused global warming representations as facts.[34]

The first UEA-sponsored investigation, called the "Scientific Assessment Panel Inquiry" headed by Lord Ronald Oxburgh, an ardent believer with strong green energy business ties, didn't assess the reliability of CRU's science either. Its scope of inquiry

---

[33] *The Climategate Inquiries*, Global Warming Policy Foundation, Andrew "Bishop Hill" Montford
[34] *ClimateGate Whitewash*, S. Fred Singer, April 14, 2010 American Thinker

was limited to reviewing papers provided to it only for evidence of deliberate misconduct. Many of those papers selected for examination by UEA were obscure, never having been challenged by critics—while others that had been criticized were not presented for review.

Lord Oxburgh's final report stated that the papers were chosen "on the advice of the Royal Society", however this was apparently untrue. In fact, many or all of those papers were reportedly selected and cleared by CRU director, Phil Jones. And contrary to strong recommendations from committee members, no public interviews were conducted, no formal notes were taken, and no recordings or transcripts of interviews were made available to the public.

The remarkably short five page Oxburgh report generously concluded that it found CRU scientists to be merely an innocent "small group of dedicated, if slightly confused researchers". It also mildly criticized IPCC for failing to cite reservations of those dedicated (but confused) researchers regarding scientific uncertainties.

Another CRU-sponsored inquiry called the "Climate Change Emails Review" headed by Sir Robert Muir-Russell hurriedly looked at more than 1,000 selected communications within a period of two and one-half weeks. Two evidence-collecting interviews were conducted with CRU staff, which the majority, including the chairman, didn't attend. No CRU critics were interviewed.

While the Muir-Russell's report concluded that the "rigour and honesty" of the CRU scientists were not in doubt, panelists admitted that the scientists' responses to "reasonable requests for information" had been "unhelpful and defensive", that "emails might have been deleted in order to make them unavailable should a subsequent request be made for them", and that there had been "a consistent pattern of failing to display a proper degree of openness, both on the part of CRU scientists and on the part of the UEA."

Did the Climate Change Emails Review accomplish the goal that Muir-Russell called for: "a concerted and sustained campaign to win hearts and minds to restore confidence in the [CRU] team's work"? The *Lancet* scientific journal's editor, Dr. Richard Horton, doesn't think so.

Testifying before the inquiry Dr. Horton said:

> The *Muir-Russell* review has rejected all claims of serious scientific misconduct. But he does identify failures, evasions, misleading actions, unjustifiable delays [in releasing information], and pervasive unhelpfulness—all of which amounts to severely sub-optimal academic practice. Climate science will never be the same again.

We can only hope that Richard Horton is right about climate science not continuing to be the same circus act it has all-too-often become. Instead it's way past time for the carnival we have witnessed to fold up its tent and get science out of show business.

See Appendix1-3 for more discussion on this topic: *The ClimateGate Whitewash-Part II: The UK's Motley CRU Reviews*, Forbes Opinions, July 5, 2011

Chapter Four: Forcing the Issues

ALTHOUGH "CLIMATE" IS generally associated with periods of at least three decades, less than one and one-half decades following mid-1970s "scientific" predictions that the next Ice Age was rapidly approaching, the media trumpeted a new and opposite alarm...a man-made global warming crisis. Previously, even the prestigious National Academy of Sciences had issued a warning that there was "a finite possibility that a serious worldwide cooling could befall the earth within the next 100 years."

The hot climate frenzy was fueled by a convergence of geopolitical circumstances. Theoretical model calculations at that time, including some at NOAA's Geophysical Fluid Dynamic Laboratory, began to indicate that substantial global warming could result from increasing CO_2 levels. Then, during a particularly hot 1988 summer in many US regions, NASA's James Hansen testified before Senator Al Gore's steamy 1988 Committee on Science, Technology and Space, that he was 99% certain that temperatures had in fact, increased due to greenhouse warming.

Since the Cold War had just ended, the Union of Concerned Scientists in search of a new cause redirected its attention from nuclear disarmament to a new "global warming threat". They issued a widely publicized statement in the *New York Times* condemning human carbon emissions as the villain.

This was also a time when Third World countries, by force of numbers, and European socialist green parties, through powers of aggressiveness, seized control of the United Nations to advance globalization goals, which emerging global warming alarm perfectly served. Accordingly, the United Nations established the Framework Convention on Climate Change (UN-FCCC) to organize conferences, along with the UN's IPCC which, prior to any studies, concluded that climate change caused by fossil burning posed a global threat.[35]

Within about half of a legitimate climate period after the earlier global cooling scares, the UN-FCCC had already determined that "climate change" was, by their definition:

> *A change of climate, which is attributed directly or indirectly to human activity, that alters the composition of the global atmosphere and which is in addition to natural climate variability observed over comparable time periods.*

Key words here are "attributed" to "human activities" which alter the "atmosphere"…greenhouse gases (CO_2 specifically).

Accordingly, when you hear references to climate change in the media these days, you can be pretty certain that it's going to discuss some observed or inevitable catastrophe attributed to "bad" greenhouse warming caused by burning evil fossil fuels. We almost never see commentary reminding us that CO_2 and warm conditions are both really great for agriculture and most Earth critters.

---

[35] United Nations Framework Convention on Climate Change: 2002

But What About Those "Observed" Human Greenhouse Influences?

The IPCC stated in its *2007 Summary for Policymaker's Report* that "Most of the observed increase in globally averaged temperature since the mid-20th century [which is very small] is very likely due to the observed increase in anthropogenic [human-caused] greenhouse gas concentrations." And there can be no doubt here that they are referring to CO_2, not water vapor, which constitutes the most important greenhouse gas of all. That's because the climate models don't know how to "observe" it, plus there aren't any good historic records to enable trends to be revealed.

Besides, unlike carbon, there is little incentive to attach much attention to anthropogenic water vapor. After all, no one has yet figured out a way to regulate or tax it.

A key problem in determining changes and influences of water vapor concentrations in the Earth's atmosphere is that they are extremely variable. Differences range by orders of magnitude in various places. Instead, alarmists sweep the problem to one side by simply calling it a CO_2 "feedback" amplification effect, always assuming that the dominant feedback is "positive" (warming) rather than "negative" (cooling). In reality, due to clouds and other factors, those feedbacks could go both ways, and no one knows for sure which direction dominates climate over the long run.

Treating water vapor as a known feedback revolves around an assumption that relative humidity is a constant, which it isn't. Since it is known to vary nearly as widely as actual water vapor concentrations, no observational evidence exists to support a CO_2 warming amplification conclusion.

But let's imagine that CO_2 is the big greenhouse culprit rather than merely a bit-player, and that its influences are predominately warming. Even if CO_2 levels were to double, it

Larry Bell

would make little difference. While the first CO_2 molecules matter a lot, successive ones have less and less effect. That's because the carbon that exists in the atmosphere now has already "soaked up" its favorite wavelengths of light, and is close to a saturation point. Those carbon molecules that follow manage to grab a bit more light from wavelengths close to favorite bands, but can't do much more…there simply aren't many left-over photons at the right wavelengths. For those of you who are mathematically inclined, that diminishing absorption rate follows a logarithmic curve.

Who Hid the Carbon Prosecuting Evidence?

Since water vapor and clouds are so complex and difficult to model, their influences are neglected in IPCC reports. What about other evidence to support an IPCC claim that "most" mid-century warming can "very likely" be attributed to human greenhouse emissions? Well, if it's there, it must be very well hidden, since direct measurements seem not to know where it is.

For example, virtually all climate models have predicted that if greenhouse gases caused warming, there is supposed to be a telltale "hot spot" in the atmosphere about 10 km above the tropics. Weather balloons (radiosondes) and satellites have scanned these regions for years, and there is no such pattern. It wasn't even there during the recent warming spell between 1979 (when satellites were first available) and 1999.

How have the committed greenhouse zealots explained this? They claim that it's there, but simply hidden by "fog in the data"…lost in the statistical "noise". Yet although radiosondes and satellites each have special limitations, their measurements show very good agreement that the "human signature" doesn't exist. Suggestions to the contrary are based upon climate model data outputs which yield a wide range of divergence and uncertainty…an example of garbage in, gospel out.

Huntington City Township
Public Library
255 West Park Drive
Huntington, IN 46750
www.huntingtonpub.lib.in.us

Scientists have long recognized that Ice Ages and brief interglacial interludes (like our current one) are caused by variations in the Earth's orbit around the sun. It is also well known that when oceans warm (in this instance due to intensification of sunlight energy), huge amounts of absorbed CO_2 are released, exactly like the out-gassing of a carbonated drink when warmed. The bottom line is that past atmospheric CO_2 is wholly controlled by the Earth's temperature and climate, not the other way around.

Also consider that atmospheric CO_2 concentrations at the end of the last Ice Age, when rapid de-glaciation occurred were less than half of today's levels. At the same time, the influence of that lower concentration would have been much greater than today due to the logarithmic absorption pattern. Therefore, the CO_2 warming amplification factor might have contributed proportionately much more influence than today, causing it to be less relevant to current circumstances.

Accurate dating of samples is very difficult and subject to large unknowns. And while carbon dioxide levels have been constantly increasing, most of all estimated warming since 1900 occurred before the mid-1940s. Despite those continuously rising CO_2 levels, Michael Mann's graph showing soaring temperatures reveals a broken hockey stick theory.

Chilling Influences of the Sun

Some recent study results linking shifts in sunspot frequency and climate changes over thousands of years suggest that the past period of flat global temperatures may be a prelude to a much longer cooling cycle. While causes behind these magnetic vacillations are uncertain, observable connections between solar and Earth climate patterns are clear. Reduced periods of sunspot activity correlate with cooler and very cold periods, with higher incidences producing opposite effects.

If a leading theory regarding why this occurs is correct, a weaker magnetic heliosphere surrounding our Solar System as evidenced by low sunspot activity permits more cosmic rays from deep space to enter Earth's protective magnetosphere and atmosphere. This increased flux of heavy electrons striking the atmosphere produces increased cloud cover, in turn reflecting more solar radiation away from Earth and back to space.

Since none of this has anything to do with human-caused atmospheric CO_2 emissions, you can bet that the UN and its Intergovernmental Panel on Climate Change are very cool on the theory and its chilling political science implications. It follows that since both the Sun and climate began changing billions of years before the Industrial Revolution, neither conditions can be blamed on fossil-fueled smokestacks and SUVs.

A notable IPCC critic is Dr. Fritz Vahrenholt, a former leader of Germany's environmental movement who also headed the renewable energy division of the country's second largest utility company. He recently co-authored a book with Dr. Sebastian Luning titled *The Cold Sun: Why the Climate Disaster Won't Happen*. And although the two don't deny that CO_2 has some warming influence, they believe the Sun plays a far greater role in the whole scheme of things.

Dr. Vahrenholt expects the world to get cooler in the future for three reasons: we are or soon will be beginning on the downward flank of the Sun's Gleissberg and Suess cycles; solar activity during the next cycle may extend our current very weak one; and ocean cycles will be in cooling phases over the next decades as well.

A research team in Sweden which analyzed patterns of solar activity at the end of the last Ice Age around 20,000-10,000 years ago concluded that changes in solar activity and their influences on climate are nothing new, especially on a regional level. An analysis of trace elements in ice cores in Greenland and cave formations from China revealed that Sweden was then covered by a thick ice sheet that stretched all the way down to northern Germany. Water contained in those frozen ice caps resulted in sea levels which were more than 100 meters lower than at present.

Furthermore, the August 2014 study report's co-author Dr. Raimund Muscheler, a lecturer in Quaternary Geology at Lund University, observes:

> Reduced solar activity could lead to colder winters in Northern Europe. This is because the Sun's UV radiation affects the atmospheric circulation.
>
> Interestingly, the same processes lead to warmer winters in Greenland, with greater snowfall and more storms.

While the Sun was exceptionally active during the 20th century, many scientists believe that this condition is now coming to an end. Although the Royal Observatory of Belgium's July average monthly sunspot count increased slightly for the sixth straight month despite a rare mid-month spotless day, solar Cycle 24 still remains to be the weakest in 100 years.

In fact, long-term indicators suggest that the next sunspot cycle will be much weaker than this one. If so, as with other extended periods of inactivity as occurred during Cycles 3, 4, and 5 which marked the beginning of a "Dalton Minimum", we can expect recent years of flat global temperatures to become significantly cooler.

Dr. Habibiullo Abdussamatov, head of the Russian Academy of Sciences Pulkovo Observatory in St. Petersburg and director of the Russian segment of the International Space Station predicts that we may soon witness the coming of a new "Little Ice Age" with a deep freeze lasting throughout this century. The last one which began in the mid-16th century killed millions in Europe. It mercifully ended soon after Washington's troops suffered brutal winter temperatures at Valley Forge in 1777 and Napoleon's bitterly cold 1812 retreat from Russia.

Whether present cooling continues or not, is there any reason at all to panic? No, and by the same token if, for any reason, global warming resumes as it probably will along with following intermittent cool-downs, let's remember to be grateful for the good times we now have the great fortune to enjoy.

See Appendix 1-4 for more discussion regarding the influence of the Sun on climate: *Sorry, With Global Warming it's the Sun, Stupid*, Larry Bell, Forbes Opinions, September 20, 2011

Chapter Five: Political Science

THE TERM "CLIMATE" is typically associated with annual world-wide average temperature records measured over at least three decades. Yet global warming observed less than two decades after many scientists had predicted a global cooling crisis prompted the United Nations to organize its IPCC and to convene a continuing series of international conferences purportedly aimed at preventing an impending catastrophe. Virtually from the beginning, they had already attributed the "crisis" to human fossil-fuel carbon emissions.

Opening remarks offered by Maurice Strong who organized the first UN Earth Climate Summit (1992) in Rio de Janeiro, Brazil revealed the real goal:

> We may get to the point where the only way of saving the world will be for industrialized civilization to collapse. Isn't it our responsibility to bring this about? [36]

Former US Senator Timothy Wirth (D-CO) who later represented the Clinton-Gore administration as US undersecretary of state for global issues had agreed, telling *The*

---

[36] *Greens Real Target: US Economy*, Investor's Business Daily, December 8, 2009

National Journal's Rochelle Stanford back in August 1988:

> *We have got to ride the global warming issue. Even if the theory of global warming is wrong, we will be doing the right thing in terms of economic policy and environmental policy.*

Wirth now heads the UN Foundation which lobbies for hundreds of billions of US taxpayer dollars to help underdeveloped countries fight climate change.[37]

Also speaking at the Rio conference, Deputy Assistant of State Richard Benedick, who then headed the policy divisions of the US State Department said:

> *A global warming treaty [Kyoto] must be implemented even if there is no scientific evidence to back the [enhanced] greenhouse effect.*[38]

In 1988, former Canadian Minister of the Environment, told editors and reporters of the *Calgary Herald*:

> *No matter if the science of global warming is all phony...climate change [provides] the greatest opportunity to bring about justice and equality in the world.*[39]

See also Appendix 1-5: *In Their Own Words: Climate Alarmists Debunk*

---

[37] National Center Dossier

[38] Zbigniew Jaworowski, *CO₂: The Greatest Scientific Scandal of Our Time*, EIR Science, March 16, 2007, p.3, 16

[39] Quoted by Terence Corcoran, *Global Warming: The Real Agenda*, Financial Post, 26 December 1998, the Calgary Herald, December, 14, 1998

Their 'Science', Larry Bell, Forbes Opinions, February 5, 2013

In 1996, former Soviet Union President Mikhail Gorbachev emphasized the importance of using climate alarmism to advance socialist Marxist objectives:

> *The threat of environmental crisis will be the international disaster key to unlock the New World Order.* [40]

Speaking at the 2000 UN Conference on Climate Change in the Hague, former President Jacques Chirac of France explained why the IPCC's climate initiative supported a key Western European Kyoto Protocol objective:

> *For the first time, humanity is instituting a genuine instrument of global governance, one that should find a place within the World Environmental Organization which France and the European Union would like to see established.*

An Intergovernmental Political Climate Crisis (IPCC)

IPCC official Ottmar Edenhofer, speaking in November 2010, advised that: "…one has to free oneself from the illusion that international climate policy is environmental policy. Instead, climate change policy is about how we redistribute de facto the world's wealth…" [41]

Most people seem to be unaware that although IPCC is broadly represented to the public as the top authority on climate

---

[40] Sovereign Independent

[41] Climate *Talks or Wealth Redistribution Talks?*, Nicolas Loris, November19, 2010, Heritage.org

matters, the organization doesn't actually carry out any original climate research at all. Instead, it simply issues assessments based upon supposedly independent surveys of published research. However, some of the most influential conclusions summarized in its reports have neither been based upon truly independent research, nor properly vetted through accepted peer-review processes.

While it should be recognized that most of the many scientific reviewers are indeed dedicated and competent people who take their work very seriously, few of them have much if any influence over final conclusions that the public hears about. Instead, the huge compilations they prepare go through international bureaucratic reviews, where political appointees dissect them, line by line, to glean the best stuff that typically supports what IPCC wanted to say in the first place. These cherry-picked items are then assembled, condensed and highlighted in the *Summaries for Policymakers* which are calibrated to get prime-time and front page attention.[42]

IPCC's 1996 report used selective data, a doctored graph, and featured changes in text that were made after the reviewing scientists approved it and before it was printed. The many irregularities provoked Dr. Frederick Seitz, a world-famous physicist and former president of the US National Academy of Sciences, the American Physical Society, and Rockefeller University, to write (in August 1996) in the *Wall Street Journal*:

> *I have never witnessed a more disturbing corruption of the peer review process than events that led to this IPCC report."*[43]

But just in case you might have lingering illusions regarding

---

[42] Roy W. Spencer, *Climate Confusion*, Encounter Books, 2008, page 147

[43] Jonathan Schell, *Our Fragile Earth*, Discover (October 1989); page 44

IPCC's scientific objectivity, let's begin with two different views by some of their same researchers that are reported in the same year regarding whether there is a discernible human influence on global climate.

First, taken from a 1996 IPCC report summary written by B.D. Santer, T.M.L Wigley, T.P. Barnett, and E. Anyamba:

> ...there is evidence of an emerging pattern of climate response to forcings by greenhouse gases and sulphate aerosols...from geographical, seasonal and vertical patterns of temperature change...These results point towards human influence on climate.

Then, a 1996 publication *The Holocene,* by T.P. Barnett, B.D. Santer, P.D. Jones, R.S. Bradley and K.R. Briffa, says this:

> Estimates of...natural variability are critical to the problem of detecting an anthropogenic [human] signal...We have estimated the spectrum...from paleo-temperature proxies and compared it with...general [climate] circulation models...none of the three estimates of the natural variability spectrum agree with each other...Until...resolved, it will be hard to say, with confidence, that an anthropogenic climate signal has or has not been detected.[44]

In other words, these guys, several of whom you will hear from later, can't say with confidence whether or not humans have had any influence at all...or even if so, whether it has caused warming or cooling!

For a bit of political science history on this matter, it's

---

[44] *Hot Talk: Cold Science,* Page 8, S. Fred Singer, The Independent Institute

important to remember that such IPCC statements typically follow a series of drafts that are edited to become increasingly media-worthy. For example, the original text of an April 2000 Third Assessment Report (TAR) draft stated, "There has been a discernible human influence on global climate." That was followed by an October version that concluded, "It is likely that increasing concentrations of anthropogenic greenhouse gases have contributed significantly to observed warming over the past 50 years.

Then in the final official summary, the language was toughened up even more:

> Most of the observed warming over the past 50 years is likely to have been due to the increase in greenhouse gas concentrations.

When the UN Environment Programme's spokesman, Tim Higham, was asked by *New Scientist* about the scientific background for this change, his answer was honest:

> There was no new science, but the scientists wanted to present a clear and strong message to policymakers.[45][46]

Kevin Trenberth, a lead author of 2001 and 2007 IPCC report chapters, writing in a 2007 *Predictions of Climate* blog appearing in the science journal *Nature.com,* admitted:

> None of the models used by the IPCC are initialized to the observed state and none of the climate states in the models correspond even remotely to the current observed

---

[45] IPCC 1996A:5, *Are Human Activities Contributing to Climate Change?*
[46] IPCC 2001D6, *A Report of Working Group I of the Intergovernmental Panel on Climate Change*

state.[47]

Christopher Landsea, a top expert on the subject of cyclones, became astounded and perplexed when he was informed that Kevin Trenberth, a lead IPCC report author, had participated in a press conference attributing a deadly 2004 Florida storm season to global warming. As news releases headlined the event:

> *Experts warn that global warming [is] likely to continue spurring more outbreaks of intense activity.*

Since IPCC studies released in 1995 and 2001 had found no evidence of a global warming-hurricane link, and there was no new analysis to suggest otherwise, Dr. Landsea wrote to leading IPCC officials imploring:

> *What scientific, refereed publications substantiate these pronouncements? What studies alluded to have shown a connection between observed warming trends on Earth and long-term trends of cyclone activity?*

Receiving no replies, he then requested assurance that the 2007 report would present true science, saying: "[Dr. Trenberth] seems to have come to a conclusion that global warming has altered hurricane activity, and has already stated so. This does not reflect consensus within the hurricane research community." After that assurance didn't come, Landsea, an invited author, resigned from the 2007 report activity and issued an open letter presenting his reasons.[48]

---

[47] Trenberth quote, K. Trenberth, *Predictions of Climate*, Climate Feedback: The Climate Change Blog, June 2007

[48] Lawrence Solomon, *The Deniers, Part III—The Hurricane Expert Who Stood Up To UN Junk Science*, National Post, February 2, 2007

For more discussion regarding candid alarmist statements see Appendix 1-6: *In Their Own Words: Climate Alarmists Debunk their 'Science'*, Larry Bell, Forbes Opinions, February 5, 2013

IPCC in a Stew About How They Cooked Their Latest Climate Books

Even the IPCC finally acknowledges in their 2013 *AR-5 Summary for Policymakers Report* that their previous estimates of "climate sensitivity" to greenhouse gases they reported in 2007 were significantly exaggerated. This has to do with projections of the amount global temperatures will increase if atmospheric CO_2 concentrations were to double.

Yes, IPCC finally admits in an obscure footnote that:

> *No best estimate for equilibrium climate sensitivity can now be given because of a lack of agreement on values across assessed lines of evidence and studies.*

But wait a minute! Wasn't that climate catastrophe "Earth at the tipping-point" alarm stuff supposed to be about a fossil-fueled CO_2 menace? After all, isn't climate sensitivity supposed to be one of the most important parameters because it determines how much warming...or cooling...we can expect? And now they're saying that they really aren't sure where or how much that dastardly greenhouse gas matters?

Based upon their models their most recent global temperature estimates project an average 3°C rise with a doubling of atmospheric CO_2 concentrations, while real world observations indicate between 1.5°C and 2°C. So maybe you got'ta wonder...If the models warmed up a lot more than the observed climate, then which of them is broke?

As former Professor of Meteorology Dr. Richard Lindzen at

MIT's Department of Earth, Atmospheric and Planetary Sciences observed:

> *The latest [2013] IPCC report truly sank to the level of hilarious incoherence—it is quite amazing to see the contortions the IPCC has to go through in order to keep the international climate agenda going.*

Also—buried in a final draft of the 2013 *Summary for Policymakers*—the writers summed up an obvious IPCC dilemma regarding their spectacular global temperature prediction failures, stating that:

> *Models do not generally reproduce the observed reduction in surface warming trend over the last 10-15 years.*

However that statement disappeared in the final version, saying instead that the difference between simulated and observed trends could be caused by some combination of (a) internal climate variability (Mother Nature); (b) missing or incorrect radiative forcing; and (c) model response error.

Still, the banner claim of this newest release is that "Human influence is extremely likely to be the dominant cause of observed warming since the middle of the last century." Remarkably, this "extremely likely" was ratcheted up from a "very likely" they claimed in their 2007 report. One can only wonder how they have become more confident that at least more than half of the temperature rise since the mid-20th century has been caused by greenhouse gas emissions, when at the same time they are less certain about climate sensitivity to CO_2.

When I asked my friend Dr. Vincent Gray, who has served as an expert reviewer for all five of the IPCC's reports, to comment on AR-5, he wasn't one bit impressed. He told me that it demonstrates IPCC believes "the public will believe almost anything that is represented as being agreed by 'scientists', provided that you have enough of them and they are backed up by the requisite number of celebrities and public figures."

Dr. Gray continued:

> Unfortunately for their message, there is no evidence that human-emitted greenhouse gases have a harmful influence on the climate. So it becomes necessary to use spin, distortion, deception and even fabrication to cover up this absence of evidence with a collective assertion of belief in their cause to an increased level of certainty. In the end they must rely merely upon collective opinions within their selected ranks, of which they once again claim high levels of certainty.

Regarding their ever increasing claims of confidence, Dr. Gray pointed out that IPCC has run into some "big trouble":

> They have been going long enough for it to become obvious that their models do not work. Even when real observed climate developments contradict their previous predictions making it obvious that their simulation models don't work, they still find it necessary to raise their confidence levels with each subsequent report. It's really crazy, but they seem to get away with it in the mainstream media.

As Dr. Gray points out:

> This latest report is infested with claims that almost everything that serves their alarmist messaging is 'very likely', a term which indicates 95% certainty. I guess they feel a need to leave that last 5% of uncertainty, just in case one day they will have to swallow their words. All

of this demonstrates that their models, and the estimates of "uncertainty" that are based upon them, are virtually useless.

When I asked Vincent what is it about the processes and products he found most disturbing from a scientific perspective he replied:

It really has little to do with objective science at all. It is much more about political spin aimed at highlighting preconceived attention-grabbing hyperbole for release to the media in Summary for Policymakers reports. Non-scientific government appointees actually control the entire report to make sure it stays exactly with the politically-controlled message they can all agree with.

In Their Own Alarming Words of Desperation

It certainly wasn't as if key IPCC authors weren't looking for possible evidence of an AGW-hurricane link they could publish. A ClimateGate note from CRU Director Dr. Phil Jones to Trenberth:

Kevin, Seems that this potential Nature [journal] paper may be worth citing, if it does say that GW [global warming] is having an effect on TC [tropical cyclone] activity.

Jones also wanted to make sure that people who supported this connection be represented in IPCC reviews:

Getting people we know and trust [into IPCC] is vital— hence my comment about the tornadoes group.

Raymond Bradley, co-author of Michael Mann's infamously

Larry Bell

flawed hockey stick paper which was featured in influential IPCC reports, took issue with another article jointly published by Mann and Phil Jones, stating:

> I'm sure you agree—the Mann/Jones GRL [Geophysical Research Letters] paper was truly pathetic and should never have been published. I don't want to be associated with that 2000 year reconstruction.

Trenberth associate Tom Wigley of the National Center for Atmospheric Research wrote:

> Mike, the Figure you sent is very deceptive...there have been a number of dishonest presentations of model results by individual authors and by IPCC...

Wigley and Trenberth suggested in another email to Mann:

> If you think that [Yale professor James] Saiers is in the greenhouse skeptics camp, then, if we can find documentary evidence of this, we could go through official [American Geophysical Union] channels to get him ousted [as editor-in-chief of the Geophysical Research Letters journal].

A July 2004 communication from Phil Jones to Michael Mann referred to two papers recently published in *Climate Research* with a *"HIGHLY CONFIDENTIAL"* subject line observed:

> I can't see either of these papers being in the next IPCC report. Kevin [Trenberth] and I will keep them out somehow—even if we have to redefine what the peer review literature is.

A June 4, 2003 email from Keith Briffa to fellow tree ring researcher Edward Cook at the Lamont-Doherty Earth Observatory in New York stated:

> *I got a paper to review (submitted to the Journal of Agricultural, Biological and Environmental Sciences), written by a Korean guy and someone from Berkeley, that claims that the method of reconstruction that we use in dendroclimatology (reverse regression) is wrong, biased, lousy, horrible, etc...If published as is, this paper could really do some damage...It won't be easy to dismiss out of hand as the math appears to be correct theoretically...I am really sorry but I have to nag about that review— Confidentially, I now need a hard and if required extensive case for rejecting.*[49][50]

Dr. Tom Crowley, a key member of Michael Mann's global warming hockey team, wrote, "I am not convinced that the 'truth' is always worth reaching if it is at the cost of damaged personal relationships."

Several email exchanges reveal that certain researchers believed well-intentioned ideology trumped objective science. Jonathan Overpeck, a coordinating lead IPCC report author, suggested:

> *The trick may be to decide on the main message and use that to guid[e] what's included and what is left out.*

Phil Jones wrote:

---

[49] Steven Mosher and Thomas Fuller, *Climategate: The Crutape Letters, Vol 1*, CreateSpace, 2010

[50] *When Scientists Become Politicians*, Investors Business Journal, December 1, 2009

Basic problem is that all models are wrong—not got enough middle and low level clouds....what he [Zwiers] has done comes to a different conclusion than Caspar and Gene! I reckon this can be saved by careful wording.

Writing to Jones, Dr. Peter Thorne of the UK Met Office advised caution, saying:

Observations do not show rising temperatures throughout the tropical troposphere unless you accept one single study and approach and discount a wealth of others. This is just downright dangerous. We need to communicate the uncertainty and be honest. Phil, hopefully we can find time to discuss these further if necessary...

In another email, Thorne stated:

I also think the science is being manipulated to put a political spin on it which for all our sakes might not be too clever in the long run.[51]

Former Alarmists Admit Buyer's Remorse

Some leading voices in the Global Warming Gospel Choir are now abandoning the old climate crisis hymnal. One is James Lovelock, the father of the "Gaia" theory that the entire Earth is a single living system who predicted that continued human CO_2 emissions will bring about climate calamity. In 2006 he claimed:

Before this century is over billions of us will die and the

---

[51] Sources (Email Quotations): *Climategate 2: The Scandal Continues*, Myron Ebell, CEI Press Release

few breeding pairs of people that survive will be in the Arctic where climate remains tolerable.

Time magazine featured Lovelock as one of 13 "Heroes of the Environment" in a 2007 article (along with Al Gore, Mikhail Gorbachev and Robert Redford).

More recently, however, he has obviously cooled on global warming as a crisis, admitting to *MSNBC* that he overstated the case and now acknowledges that:

> ...we don't know what the climate is doing. We thought we knew 20 years ago. That led to some alarmist books...mine included...because it looked clear cut...but it hasn't happened.

Lovelock pointed to Al Gore's *An Inconvenient Truth* and Tim Flannery's *The Weather Makers* as other alarmist publications.

The 92-year-old Lovelock went on to note, "...the climate is doing its usual tricks...there's nothing much happening yet even though we were supposed to be halfway toward a frying world now." He added, "The world has not warmed up very much since the millennium. Twelve years is a reasonable time." Yet the temperature "has stayed almost constant, whereas it should have been rising...carbon dioxide has been rising, no question about that."[52]

Former German Green Movement leader and Socialist Dr. Fritz Vahrenholt raised a man-made blizzard of criticism by charging IPCC with gross incompetence and dishonesty, most particularly regarding fear-mongering exaggeration of known climate influence of human CO_2 emissions. His distrust of the IPCC's objectivity and veracity first took root when he became an

---

[52] *Environmentalist Icon Says He Overstated Climate Change*, Investor's Business Journal, April 25, 2012

expert reviewer for their report on renewable energy. After
discovering numerous errors, he reported those inaccuracies to
IPCC officials, only to have them simply brushed aside. Stunned
by this, he asked himself, "Is this the way they approached climate
assessment reports?" He came to wonder: "...if the other IPCC
reports on climate change were similarly sloppy."

This concern prompted Vahrenholt to dig into the IPCC's
2007 climate report, and he was horrified by what he found. On
top of discovering numerous factual errors, there were issues
involving 10 years of stagnant temperatures, failed predictions,
ClimateGate emails, and informative discussions with dozens of
other elite skeptical scientists.

Vahrenholt concludes in an interview which appeared in the
German news publication *Bild* that:

> ...*IPCC decision-makers are fighting tooth and nail
> against accepting the roles of the oceans, sun, and soot.
> Accordingly, IPCC models are completely out of whack.
> The facts need to be discussed sensibly and scientifically,
> without first deciding on the results.* [53] [54]

Vahrenholt isn't the only significant German scientist to find that
IPCC's global warming projections are exaggerated. Another is
Hans Joachim Schellnhuber, the director of the Potsdam Institute
for Climate Impact Research who serves as the German
government's climate protection advisor. Schellnhuber
coauthored a paper refuting reliability of Global Climate Models
upon which their alarmist 2001 projections were based.

The study compared measured versus model-simulated

[53] *The Compelling Case Against Ed Davey*, Melanie Phillips, February 7, 2012, The Daily Mail.com
[54] *A Top German Environmentalist Cools On Global Warming*, Larry Bell, February 14, 2012, Forbes,

temperature trends at six global sites according to two different scenarios; one with greenhouse gas influence plus aerosol influences, and the other with greenhouse temperature influences only. Results showed that while both scenarios failed to reproduce observed temperature recordings, the one using only greenhouse influences demonstrated the greatest deviation from reality: "...where the [greenhouse gas scenario] trends are clearly overestimated."[55]

Schellnhuber admitted in a speech to agricultural experts that: "warmer temperatures and high CO_2 concentrations in the air could very well lead to higher agricultural yields."[56]

Dr. Patrick Moore, a cofounder of Greenpeace, quit the activist environmental organization in 1986 after it strayed from objective science and took a sharp turn to the political left. Testifying on February 25, 2014 before the Senate Environmental and Public Works Committee's Subcommittee on Oversight, he took issue with the IPCC's claim that "Since the mid-20th century it is 'extremely likely' that human influence has been the dominant cause of the observed warming."

Dr. Moore pointed out that "There is no scientific proof that human emissions of carbon dioxide are the dominant cause of the minor warming of the Earth's atmosphere over the past 100 years", arguing that "perhaps the simplest way to expose the fallacy of extreme certainty is to look at the historical record." He told the committee, "When modern life evolved over 500 million years ago, CO_2 was more than 10 times higher than today, yet life flourished at this time. Then an Ice Age occurred 450 million years ago when carbon dioxide was 10 times higher than today.

Moore noted that the increase in temperature between 1910

---

[55] Physics Review Papers

[56] *German Fear of Warming Plummets...Yet-To-Be-Published Skeptic Book Climbs To Amazon.de No.4!*, P. Gosselin, January 30, 2012, NoTricksZone

and 1940 was virtually identical to the increase between 1970 and 2000. Yet the IPCC does not attribute the increase from 1910-1942 to human influence.

Why then, he asks, does the IPCC believe that a virtually identical increase in temperature after 1950 is caused mainly by human influence, when it has no explanation for nearly identical increase from 1910 to 1940?

Moore also observed that there is no reason to believe that a warmer climate would be anything but beneficial for humans and the majority of other species. On the other hand, there is ample reason to believe that a sharp cooling of the climate would bring disastrous results for human civilization.

A Public Climate Change Threatens Alarmists

The release of scandalous email exchanges among IPCC scientists has taken large tolls in public climate opinion polls. A majority of Americans nationwide acknowledge that there is significant disagreement about global warming in the scientific community. Most responders even go even further, believing that some scientists have falsified data to support their own beliefs.

An August 2011 *Rasmussen Reports* national telephone survey of American Adults showed that 69% said it is at least "somewhat likely" that some scientists have falsified research data in order to support their own theories and beliefs, including 40% who said this is "very likely". (The number who said it's likely is up 10 points since December 2009.) And while Republicans and adults not affiliated with either major political party felt stronger than Democrats that some scientists have falsified data to support their global warming theories, 51% of the Democrats also agreed.

As for "scientific consensus", 57% of those surveyed believed there is significant disagreement within the scientific community on global warming. This was up five points from late 2009.[57]

Rapidly growing public skepticism in the US and abroad about the veracity climate calamity claims is now putting alarmists on the defensive. As Paul Ehrlich at Stanford University reported in a March 2010 *Nature* journal editorial, this has his colleagues in big sweats about how to counter a barrage of challenges:

> *Everyone is scared shitless, but they don't know what to do."* [58]

And speaking of scary, Ehrlich is best known for his 1968 doom and gloom book, *The Population Bomb*, which predicted that a worldwide crisis in food supply and natural resource availability would lead to major famines and economic failures by 1900. In another book titled *The Machinery of Nature*, he predicted that carbon dioxide-induced famines might kill as many as a billion people by 2000.

Ehrlich's claims were based upon a theory advanced by none other than John Holdren, who is now serving as the Obama administration's Science Czar. The central premise was that human CO_2 emissions would produce a climate catastrophe in which global warming would cause global cooling...resulting in widespread agricultural disaster. Holdren's theory was that the warm temperatures might speed up air circulation patterns to bring Arctic cold farther south, and Antarctic cold farther north.[59]

Yup, in other words, they were worried about a global warming-induced ice age.

During a March 2011 Senate Environment and Public Works Committee hearing, Senator James Inhofe (R-OK) cited a 1971

---

[57] *69% Say It's Likely Scientists Have Falsified Global Warming Research*, August 3, 2011, Rasmussen Reports
[58] *Climate of Fear*, Nature Editorial, March 11, 2010
[59] *The Machinery of Nature*, Simon & Schuster, New York, 1986, p. 274

study where Holdren wrote, "The effects of a new ice age on agriculture and the supportability of large human populations scarcely need elaboration here." Senator Inhofe then turned to Senator Boxer (D-CA) Boxer, and stated, "So even the president's people are agreed with me, Madam Chairwoman!"

Accordingly, even if Holdren's calamitous prediction never caught the planet by storm, let's at least give him some credit for realizing that global warming is a lot more life-friendly than the opposite. Isn't it unfortunate that most members of his overheated chorus aren't in concert with this brutally cold fact? [60]

The Costs of Ideology Masquerading as Science

As Greenpeace co-founder Peter Moore observed on Fox Business News in January 2011: "We do not have any scientific proof that we are the cause of the global warming that has occurred in the last 200 years…The alarmism is driving us through scare tactics to adopt energy policies that are going to create a huge amount of energy poverty among the poor people. It's not good for people and it's not good for the environment…In a warmer world we can produce more food."[61]

When Moore was asked who is responsible for promoting unwarranted climate fear and what their motives are, he said:

A powerful convergence of interests. Scientists seeking grant money, media seeking headlines, universities seeking huge grants from major institutions, foundations, environmental groups, politicians wanting to make it look like they are saving future generations. And all of

---

[60] *Senators Spar During Hearing Over Alleged 1970s Cooling Consensus*, Daily Caller, March 3, 2011
[61] *Environment Greenpeace Founder Questions Man-Made Global Warming*, Jonathan M. Seidl, January 20, 2011, The Blaze

these people have converged on this issue.

The US Government Accounting Office (GAO) reports that federal climate spending has increased from $4.6 billion in 2003 to $8.8 billion in 2010 (a total $106.7 billion over that period). This doesn't include $79 billion more spent for climate change technology research, tax breaks for "green energy", foreign aid to help other countries address "climate problems"; another $16.1 billion since 1993 in federal revenue losses due to green energy subsidies; or still another $26 billion earmarked for climate change programs and related activities in the 2009 "Stimulus Bill".[62]

Virtually all of this is based upon unfounded representations that we are experiencing a known human-caused climate crisis, a claim based upon speculative theories, contrived data and totally unproven modeling predictions. And what redemptive solutions are urgently implored? We must give lots of money to the UN to redistribute; abandon fossil fuel use in favor of heavily subsidized but assuredly abundant, "free", and "renewable" alternatives; and expand federal government growth, regulatory powers, and crony capitalist-enriched political campaign coffers.

It is way past time to realize that none of this is really about protecting the planet from man-made climate change. It never was.

---

[62] *The Alarming Cost of Climate Change Hysteria*, Larry Bell, Forbes Opinions, August 23, 2011

Chapter Six: Climate as Religion

GLOBAL WARMING, AKA climate change, has become a religious mantra, a call for action in a crusade against larger evils we have perpetrated against nature, a punishment for our sins. Author Michael Crichton articulated the essence of this creed in a 2003 speech whereby:

> There's an initial Eden, a paradise, a state of grace and unity with Nature; there's a fall from grace into a state of pollution as a result from eating from the tree of knowledge; and as a result of our actions, there is a judgment day coming for all of us. We are energy sinners, doomed to die, unless we seek salvation, which is now called sustainability. Sustainability is salvation in the church of the environment, just as organic food is its communion, that pesticide-free wafer that the right people with the right beliefs imbibe.[63]

Let's recognize that Mr. Crichton was not arguing against the importance of living more environmentally responsible lives that apply resources in sustainable ways, or against the central role that

---

[63] Roy W. Spencer, *Climate Confusion*, (Encounter Books, 2008), 98-99

religion plays in guiding most of us, whether we subscribe to a particular orthodoxy or not. But that idyllic view of an Eden in the "good old days" before industrialization and modern technology wrecked everything warrants some objective reflection.

For Better and Worse, Climate Changes with No Help from Us

Realities going back a few hundred years and more reveal a different picture; one displaying widespread poverty, starvation, disease and hardship. Yes, throughout human history, people have had to adapt to climate changes—some long, some severe, and often unpredictable. They have blamed themselves for bad seasons, believing they had invoked the displeasure of the gods through a large variety of offenses. High priests of doom told them so, extracting oaths of fealty and offerings of penance for promised interventions on their behalf. In this regard, at least for some, it seems little has changed. That penance today comes at a very high cost...our present and future national economy.

Nobel Physics laureate Ivar Giaever has called global warming a "new religion". Its temple is built on grounds of faith rather than scientific foundations. Climate change is not Mother Nature's retribution for human audacity to multiply and survive, any more than a tornado that destroys a particular church is God's retribution for belonging to the "wrong" congregation. Get over it! It's not all about us! [64]

Climate changes and shorter-term weather events are the way nature balances itself, move heat and moisture around, and provide motivations for species to evolve. CO_2 is a small but nonetheless important part of the system. Without it life would

---

[64] *Ivar Giaever: Scientists Debunk Global Warming*, Geoff Metcalf, Newsmax.com, December 15, 2008

not exist at all. No polar bears, no penguins, no coral reefs—and certainly no rain forests that directly breathe in lots of the stuff. Don't call it "pollution". At least show it a little respect!

Global warming has been effectively marketed by doomspeakers because it provides really exciting visual impressions: icebergs calving, polar bears exhausted from swimming, and such. Endless "authorities" will back up these images with scary prophesies regarding just how bad things are likely to get based upon speculative theories and unproven computer models offered as articles of faith.

Religious history is replete with stories of floods, from Noah to Gilgamesh. But melting and freezing patterns are far too complex and regional to be predicted by models. In addition, accurate satellite records only recently became available. It's also important to note that higher and lower atmospheric CO_2 concentration trends have tended to follow, not lead, temperature changes which have occurred for a large variety and combination of natural reasons.

Earth's Apocalyptical "Tipping Point" Probably Isn't So Near After All

Knee-jerk responses to alarmist forecasts make for great media, but do so at the expense of good science. If claims that continuous Greenland melting accelerations were correct, even at previously measured advancing rates (and precluding intervention of another potentially overdue Ice Age), it would take thousands of years to significantly affect sea levels.

Anyone with even a modicum of knowledge about climate history recognizes that the arctic experiences substantial climate swings about every 60-70 years due to entirely natural ocean oscillation cycles. It was very much warmer 1,000 years ago when Eric the Red and his band of Icelandic Viking settlers raised livestock on Greenland's coastal grasslands. I certainly didn't see any sheep or goats during my year there in 1959-60 as a military air traffic controller when temperatures reached 60 degrees below zero.

Global warming zealots have launched an aggressive jihad against those they brand as "deniers", often asserting presumed fossil energy affiliations to those who don't buy into their hysterical pronouncements. A central tactic is to conflate rejection of unfounded human-caused climate crisis alarmism with an absurd denial that climate changes regularly occur—or that human activities may have some influence (however incalculably miniscule, either with respect to warming or cooling, those effects may be). They would also have us believe that periods of global warming are assuredly "bad", and have even caused recent global cooling! In other words, *all* climate change is bad.

Sins of Progress and Prosperity

In the church of climate change, most or all unfortunate events that occur are attributable to human causation. Eco-elitists seize upon this dogma to argue that economic growth, promulgated by spurious corporate interests, is the enemy of the environment. They overlook the fact that that economic progress yields technological innovation and prosperity essential to support more resourceful, cleaner and healthier lifestyles. A return to small, self-sufficient, agrarian communal societies of our ancestors is no longer practical or desirable, either for us or other creatures that share and depend upon common ecosystems.

Whether or not we subscribe to a particular orthodoxy, religion plays a vital, if not central role in most of our lives, guiding us to believe we are all parts of something much larger than ourselves. It provides age-old lessons that teach us the importance of taking responsibility for our actions, constantly motivating us to do better. Faith in those universal principles binds us together as stable, functioning societies.

Science also has a vital—but very different role. When purported "scientific experts" emulate spiritual prophets they overstep their bounds, and we can no longer trust them.

Section 2

Disastrous Speculations

PRESIDENT OBAMA HAS put salvation from dreaded climate catastrophes on his action agenda hot list. During his inaugural address he said:

> We will respond to the threat of climate change, knowing that the failure to do so would betray our children and future generations.

He went on to shame anyone who disagrees with this assessment, saying:

> Some may still deny the overwhelming judgment of science, but none can avoid the devastating impact of raging fires and crippling drought and powerful storms.

This sort of scary presidential prognostication isn't new. He previously emphasized at the Democratic National Convention that global warming was "not a hoax", referred to recent droughts and floods as "a threat to our children's future", and pledged to make the climate a second term priority.

Larry Bell

Not one to let that perfectly good crisis go to waste, Secretary of State Kerry has been even more dramatic than his boss, describing climate change as the world's "most fearsome weapon of mass destruction". Speaking at a February 2014 press conference in a US Embassy-run American Center held at a shopping mall in Jakarta, Indonesia Kerry also referred to those who don't subscribe to that hell-in-a-hand-basket global warming apocalypse as "Flat-Earthers".

Kerry told the audience:

> We should not allow a tiny minority of shoddy scientists and science and extreme ideologues to compete with scientific facts. The science is unequivocal, and those who refuse to believe it are simply burying their heads in the sand. We don't have time for a meeting anywhere of the Flat Earth Society.

Chapter Seven: Climate Change: it's coming to Your Neighborhood Soon!

AS TRUMPETED IN *The Third National Climate Assessment* (NCA) released in May 2014, "Climate change, once considered an issue for a distant future, has moved firmly into the present."

Even if you somehow imagined that dramatic climate changes haven't been going on throughout our planet's history, have no doubt that this ain't just any old ordinary climate change conditions they're talking about. Nope, you can be certain that it's about global warming influences that our fossil-fueled smokestacks and SUVs are causing. And you can also bet your bippy that they're all considered to be bad.

As Dr. Judith Curry, chairwoman of the Georgia Institute of Technology School of Earth and Atmospheric Sciences observes:

> *The report effectively implies that there is no climate change other than what is caused by humans, and that extreme weather events are equivalent to climate change...Worse yet, is the spin being put on this by the Obama administration.*

Dr. Roy Spencer, principal research scientist at the University of Alabama-Huntsville takes this "spin" criticism a bit farther. He states that part of the report "is just simply made up. There is no fingerprint of human-caused versus naturally-caused climate change."

Meteorologist and *Weather Channel* Co-founder John Coleman doesn't have a very high opinion of the NCA either. He refers to the report as a "litany of doom", calling it a "total distortion of the data and an agenda-driven, destructive episode of bad science gone berserk."

The 829 page report's oft-quoted banner headline is that "extreme weather events with links to climate change have become more frequent and/or intense". The summary "overview" asserts that evidence confirming that this trend is already "disrupting people's lives" tells an "unambiguous story".

Well, at least that's the big message until you get to the fine print in the body of the report which acknowledges that, oops, maybe that evidence is more ambiguous than they originally headlined. Here they admit "trends in severe storms, including the intensity and frequency of tornados, hail, and damaging thunderstorm winds, are uncertain and are being studied intensively". The report also observes "There has been no universal trend in the overall extent of drought across the continental US since 1900".

In fact many of those claims are more than uncertain. They are most certainly wrong.

Extreme Nonsense

Meteorologist Joe Bastardi argues that there is really nothing very extreme about current climate and weather conditions which are similar to major events that occurred on a regional scale in the early 1950s. He points out that such changing climate and fluctuating weather consequences are driven primarily by natural changes in solar cycles, ocean temperatures and "stochastic events" such as volcanoes. The first two occur on various long-term cycles; decades and centuries with the Sun, and decades for the oceans. The stochastic events are random wild cards.

The early 1950s was the last time the Pacific Ocean shifted its temperature phase from warm to cold when the Atlantic was in a warm phase, and globally, the Earth's temperatures have fallen about .05C in the last four years. The European and Far East winters also look very similar now to those in the 1950s. Alaska has once again turned much colder, just as it did then when the Pacific temperatures cooled and sea ice expanded.

Here in the US, a drop in tropical Pacific temperatures causes less moisture to be present in the atmosphere than when that ocean is in its warming state. That causes conditions to be drier, especially near and east of the Rockies, as well as in the Deep South. This is when we see hotter, drier summers; winters tend be warmer earlier, and colder later. We're also seeing colder spring temperatures caused by a multi-decadal warming temperature shift in the Atlantic which has greatest influence in the late winter and spring, forcing what is called a negative North Atlantic Oscillation (NAO).

Based upon ocean temperatures alone (not including solar influences), springs although not as extremely cold, can be expected to be more common during the next 5 to 10 years. We can also expect to witness increased tornado activity which is linked to the cold decadal Pacific shift and a cooling globe. This happens when cool air in northwest North America trying to find a pathway southeast collides with warm air coming from the south in a clash zone right in the center of our nation.

As Joe Bastardi observed to me, "Mother Nature is always searching for a balance she can never fully achieve because of the design of the system. It's not unlike Aquinas's search for the unmoved mover. We rotate around the Sun on an axis that tilts, with more land in the Northern Hemisphere than Southern Hemisphere. [Climate and] weather is a movie, so we have to keep an eye on what the director is up to."

So while some alarmists have screamed about the northern ice cap melting due to warming, a condition actually caused by the Atlantic's multi-decadal phase, the southern ice cap has increased to record levels. This is even a more impressive feat because, since being surrounded by water, it requires more cooling to freeze that ice than it does to warm cold dry air on continental surfaces that surround the Arctic Ocean. But as soon as the Atlantic goes into its cold mode, the northern ice cap will expand again as the southern ice cap shrinks. This is but one example of how the back-and-forth mechanism works.

Professor and Chair of the School of Earth and Atmospheric Sciences at the Georgia Institute of Technology, Dr. Judith Curry, also notes important contributions of 60-year Pacific and Atlantic Ocean temperature cycles, observing that they have been "insufficiently appreciated in terms of global climate". When both oceans were cold in the past, such as from 1940 to 1970, the climate cooled. The Pacific "flipped" back from a warm to a cold mode in 2008, and the Atlantic is also thought likely to flip back in the next few years.[65]

Take hurricanes for example, where a century-long trend is actually down. In 2013 the National Hurricane Center stated "There were no major hurricanes in the North Atlantic Basin for the first time since 1994. And the number of hurricanes this year was the lowest since 1982." The US is experiencing the longest period without major Category 3, 4 or 5 hurricane landfall strikes since the Civil War era. The global frequency of tropical hurricanes is now also at a historical low.

As for tornadoes, the National Oceanic and Atmospheric Administration reported that Dorothy is safer than ever back in Kansas. They tell us, "There has been little trend change in the frequency of the stronger tornadoes over the past 55 years."

---

[65] *When the Earth Refuses to Warm*, Wesley Pruden, January 31, 2012, The Washington Times

NOAA's US Climate Extremes Index of unusually hot or cold temperatures finds nothing to be alarmed about either. Five years during the past ten have recorded temperatures below the historical mean, and five have been recorded above.

But what about that catastrophic CO_2-induced global warming all those really sophisticated climate models have warned us about? Well, apparently it has taken a cool shady siesta. Even NOAA admits that there has been a "lack of significant warming at the Earth's surface in the past decade" and a pause "in global warming since 2000." As they stated, "since the turn of the century, however, the change in Earth's global mean surface temperature has been close to zero."

And those "abnormal" extreme drought and moisture conditions we have been witnessing? Well again, maybe not so much after all. Floods have been occurring since the writing of the Old Testament, and California is no stranger to droughts.

While parts of the country have indeed experienced higher than average drought/moisture conditions over the past ten years, four of those years have been below average and six have been above. Conditions during the 1930's, 40's and 50's when atmospheric CO_2 levels were higher were more extreme. And by the way, US floods haven't increased in frequency or intensity since at least 1950.

A recent study published in the letter of the journal *Nature* indicates that globally, "...there has been little change in drought over the past 60 years." And, as the UN climate panel concluded last year:

> *Some regions of the world have experienced more intense and longer droughts, in particular in southern Europe and West Africa, but in some regions droughts have become less frequent, less intense, or shorter, for example,*

Larry Bell

in central North America and northwestern Australia.[66]

Wouldn't you think that instead of urging us to build arks in our back yards, the Obama administration would wish to take some credit for all this good news? Perhaps remember that this is the president who promised on the night he won the Democratic nomination that it was "the moment when the rise of the oceans began to slow and our planet began to heal."

You got'ta hand it to him for modesty.

Climate of Fear and Foreboding

Climate alarm isn't a new thing, and has sometimes been downright chilling. Back in 1975 the New York Times warned that the present period of benign interglacial climate that has existed for the past 10,000 years, warm conditions which have occurred over only about 8% of the time over the past 700,000 years are likely to end abruptly with major cooling inevitable. Time magazine agreed, reporting "The trend shows no indication of reversing. Climatological Cassandras are becoming increasingly apprehensive, for the weather aberrations they are studying may be the harbinger of another Ice Age." [67 68 69]

Newsweek agreed, following up with an April 28, 1975 article titled *The Cooling World*. It took them 31 years to admit their mistake...sort of. Senior Editor Jerry Adler didn't really blame it on journalism. Instead, he rationalized that "Some

---

[66] *Little Change in Global Drought over the Past 60 Years*, Justin Sheffield, Eric F. Wood and Michael L. Roderick, November

[67] New York Times: *Climate Changes Called Ominous*, June 19, 1975, Harold M. Schmeck, p. 31

[68] New York Times: *Scientists Ask Why World Climate is Changing, Major Cooling May Be Ahead*, May 21, 1975, Walter Sullivan

[69] Time Magazine: *Another Ice Age*, June 24, 1974

scientists indeed thought the Earth might be cooling in the 1970s, and some laymen—even one as sophisticated and well-educated as Isaac Asimov—saw potentially dire implications for climate and food production," He also didn't apologize for the fact that those reports sold a lot of magazines.[70]

Chilling climate sensationalism has sold lots of books too. Lowell Ponte's *The Global Cooling: Has the Next Ice Age Already Begun?* was a big hit in 1976. Ponte admonished that "This cooling has already killed hundreds of thousands of people. If it continues and no strong action is taken, it will cause world famine, world chaos and world war...."

Prominent "climate experts" joined the fearsomely frigid fray. In 1977 NASA scientist S.I. Rasool, a colleague of famous global warming alarmist James Hansen, predicted that by 2020, fossil fuel dust injected into the atmosphere by man "could screen out so much sunlight that the average temperature could drop by six degrees, resulting in a buildup of new glaciers that could eventually cover huge areas." If sustained over "several years, five to ten," or so Rasool estimated, "such a temperature decrease could be sufficient to trigger an Ice Age." That conclusion was presented in a paper published in the journal *Science* that was co-authored by the late Stephen Schneider, a prominent scientist and member of the UN's IPCC.[71][72][73]

The late world-renowned atmospheric scientist Dr. Reid Bryson agreed with Rasool and Schneider, stating that "Whatever the cause of the cooling trend, its effects could be extremely serious, if not catastrophic. Scientists figure that only a 1%

---

[70] *The Cooling World*, Newsweek, Gwynne

[71] 1977 book: *The Weather Conspiracy: The Coming of the New Ice Age*—CIA Feared Global Cooling

[72] July 9, 1971, Washington Post article presented in the Washington Times, September 19, 2007

[73] *Stephen Schneider, Greenhouse Superstar*, John L. Daly

decrease in the amount of sunlight hitting the earth's surface could tip the climatic balance, and cool the planet enough to send it sliding down the road to another Ice Age within only a few hundred years."

Dr. Bryson later decided that human fossil fuel burning couldn't really be blamed for causing a climate catastrophe after all. Twenty years later in 2007 he said, "Before there were enough people to make any difference at all, two million years ago, nobody was changing the climate, yet the climate was changing, okay?" He went on to comment, "You can go outside and spit and have the same effect as doubling carbon dioxide." [74]

While many National Academy of Sciences meteorologists disagreed about the cause and extent of the 1970s cooling trend, as well as over its specific impact on local weather conditions, they were almost unanimous in the view that the trend will reduce agricultural productivity resulting in potentially catastrophic famines. Yet they also almost unanimously conceded that some of the more spectacular solutions proposed, such as melting the Arctic ice cap by covering it with black soot or diverting arctic rivers, might create problems far greater than those they solve. [75]

See also Appendix 2-1: *Leading Scientists Predict Next Ice Age Coming: Hell to Freeze Over!*, Larry Bell, Forbes Opinions, 2012

Another Scientific Meltdown

As climatologist and fellow Forbes contributor Dr. Patrick Michaels recalls "When I was going to graduate school, it was

---

[74] *The Faithful Heretic: A Wisconsin Icon Pursues Tough Questions*, The Wisconsin Energy Conservative

[75] *Scientists Considered Pouring Soot Over the Arctic in the 1970s to Help Melt the Ice—In Order to Prevent Another Ice Age*, December 17, 2009, Washingtonsblog

gospel that the Ice Age was about to start. I had trouble warming up to that one too. This (greenhouse) is not the first climate apocalypse, but it's certainly the loudest."

And we survived an earlier one before. That was more than 90 years ago when the *Associated Press* and *Washington Post* reported that fishermen, seal hunters and explorers in Norway were observing a radical change in climate and hitherto unheard-of high temperatures in the Arctic. Scarcely any sea ice had been seen as far north as 81 degrees 29 minutes, soundings to a depth of 3,100 meters indicated the Gulf Stream was unusually warm, great masses of ice had been replaced by moraines of earth and stones, and at many points, well known glaciers had disappeared.

The report went on to lament that very few seals and no white fish were being found in the Eastern Arctic, while vast shoals of herring and smelt were being encountered in the old seal fishing grounds. All of this supported a prediction that due to the ice melt the sea would rise, making most cities uninhabitable. Sound familiar? [76]

---

[76] *Arctic Ocean Warming, Icebergs Growing Scarce Washington Post Reports,* Kirk Myers, March 2, 2010, Examiner.com

Chapter Eight: Goodness Glaciers!

TWO BREATHLESSLY ANNOUNCED 2014 reports suggest that global warming is causing irreversible melting of Antarctic glaciers which will cause a catastrophic sea level rise. One of those studies conducted by a group at NASA's Jet Propulsion Laboratory and the University of California-Irvine used satellite measurements to track "six rapidly melting glaciers" between 1992 and 2011.

The other, conducted at the University of Washington, developed computer models to predict that coastal melting could cause the collapse of the West Antarctic ice sheet within 200 to 900 years. This, they claim, might cause global sea levels to rise by as much as 10 feet.

As for influences surrounding this knuckle-biting alarm, the researchers place blame on broad patterns of climate change. Included are rising regional temperatures, warming ocean currents and changing wind patterns. And while both research teams agreed that the existence of warm water might actually be part of the natural ocean system, they assert that climate change is a contributing factor in bringing the warm water in contact with the ice and causing it to melt.

But wait a minute. This isn't exactly a new development. The West Antarctic Ice Sheet (WAIS) has been melting at about its recent rate over thousands of years.

Just for the Record

A glacier at Pine Island triggered great scientific interest when JPL/UC-Irvine satellite measurements revealed it has retreated 19 miles since 2005. This fast moving ice discharger first gained lots of attention when satellite measurements revealed a retreat rate of about 1.2 kilometers per year between 1992 and 1996. A large iceberg which broke loose from the glacier caused media-trumpeted speculation that this "race to the sea" heralded the beginning of the end for the West Antarctic ice sheet.

Another satellite study reported in a 2000 issue of the *Journal of Geophysical Research* determined that the dynamics of ice thinning over the entire Pine Island drainage basin which amounted to about 1.6 meters per year between 1992 and 1999 was most likely driven by phenomena operating on time scales of hundreds to thousands of years, not by twentieth century warming.

And what net contribution to sea level rise will result if this thinning rate continues unabated? It will amount to about 6 mm—the width of a standard paper clip—each century.

Regarding dramatic shifts in the WAIS, there's nothing new about that either. A major study released by the British Antarctic Survey found that the Pine Island glacier thinned just as rapidly 8,000 years ago as it has in recent times, yet subsequently recovered.

Another BAS study published in the journal *Geophysical Research Letters* reported that WAIS thinning is "within the natural range of climate variability" over the past 300 years. It also noted that "More dramatic isotopic warming (and cooling) trends occurred in the mid-19th and 18th centuries", suggesting that the anthropogenic (human-caused) climate influences in this location can't be the driver.

Geological evidence suggests that the Antarctic coastline

which is now covered with ice was ice-free only 6,000 years ago. Much more recently in 1513 a Turkish sea captain named Piri Reis was able to chart open coastline waters, so that doesn't really tally with global warming (or cooling) as recent climate change events either.

A 1922 article which appeared in a South Australia newspaper warned:

> *We are now, it is believed, slowly approaching another warm epoch, when, if it becomes universal, affecting both hemispheres together, the ice will again melt, and the sea rise to its ancient level, submerging an enormous portion of what is now dry and thickly populated land.*

Meanwhile, as the WAIS once again thins, Antarctic sea ice continues to grow and break recent records. At the end of May 2014 that extent reached the highest level recorded since satellite measurements first began in 1979. The expanse reached nearly 13 million square kilometers, beating the previous 2010 record of 12.7 million kilometers. That new record was 10.3 percent above the 1981-2010 climatological average of 11.7 million square kilometers.

For some additional perspective, consider that the WAIS which has been experiencing modest warming contains less than 10% of the continent's total ice mass. The other 90% has been getting colder. At the current melting rate we might expect the West Antarctic Ice Sheet to be gone in another 7,000 years or so...provided that the next in a series of Ice Ages doesn't intervene as normally scheduled. Other smaller ice sheets that once existed in the Antarctic are already gone.

Scientists involved with the most recently reported studies admit that the West Antarctic Ice Sheet isn't melting due to warming air temperatures, but rather because naturally-occurring warm ocean water is being pulled to the surface by the intensification of winds that encircle the continent. They then hypothesize that those stronger winds are being influenced by human-caused global warming. Yup, we caused it anyway!

Here's something else those study reports didn't bother to mention. In 2012 some experts from the University of Aberdeen and British Antarctic Survey discovered a huge one-mile-deep rift valley about the size of the Grand Canyon located beneath the ice in West Antarctica. Since this previously-hidden ice-filled basin connects directly with the warmer ocean, they think it might constitute a major cause for much of the melting in this region.

Does Global Warming Cause Volcanoes?

It might also be worth mentioning that a chain of active volcanoes has recently been discovered under that WAIS. While it is believed that eruptions are unlikely to penetrate the 1.2 to 2-kilometer-thick overlying ice, researchers conclude that they could generate enough melt water to significantly influence ice stream flow.

See also Appendix 2-2: *W. Antarctic Ice Sheet Melt Reports Omit Non-Alarming Facts*, Larry Bell, Newsmax, May 19, 2014

A recent study published in the *Proceedings of the National Academy of Sciences* presents strong evidence that this geothermal heat from volcanic magma, not global warming, is a major factor contributing to glacial melting.

Using radar techniques to map water flows under the Thwaites Glacier in Western Antarctica, researchers at University of Texas Institute for Geophysics found the resulting heat was much more widely and less evenly distributed than previously believed, with some areas considerably hotter than others. Thwaites, one of the world's largest and most rapidly retreating glaciers, is of special interest for predicting sea level rise.

Getting a good handle on the relative influence of natural thermal conditions has proven to be very challenging. As the report states:

> *Geothermal flux is one of the most dynamically critical ice sheet boundary conditions, but is extremely difficult to constrain at the scale required to understand and predict the behavior of rapidly changing glaciers.*

Yet as UTA lead researcher David Schroeder commented in a press release, "The combination of variable sub-glacial geothermal heat flow and the interacting sub-glacial water system could threaten the stability of Thwaites Glacier in ways that we never before imagined.

Previous estimates depended extensively upon theoretical modeling. But as UTA report coauthor Don Blankenship observed:

> *It's the most complex thermal environment you might imagine. And then you plop the most critical dynamically unstable ice sheet on planet Earth in the middle of this thing, and then you try to model it. It's virtually impossible.*

See also Appendix 2-3: *Study Attributes Antarctic Glacier Melt to Volcanoes*, Larry Bell, Newsmax, June 23, 2014

Let's finally realize that West Antarctic Ice Sheet has been melting at about its recent rate over thousands of years. This condition is expected to continue until either it entirely disappears, or until such time as the next Ice Age intervenes to stop it. Let's hope that the latter circumstance doesn't occur any time soon.

Behaviors Not So Simple as Imagined

Glaciers appear to have minds of their own. A world-wide study of 246 of them conducted by Dr. Roger Braithwaite, a senior research fellow at the School of Environment and Development, University of Manchester, UK, revealed that between 1946 and 1995 some have experienced overall shrinkage while others have grown, with no common or global trend of increasing melt in recent years. Within Europe, for example, Alpine glaciers generally shrunk, Scandinavian glaciers grew, and those in the Caucasus were close to equilibrium from 1980-1995.[77]

Studies conducted at Holland's Institute for Marine and Atmospheric Research in Utrecht published in the 2008 issue of the journal *Science* based upon 17 years of satellite measurements note that Greenland's recent melt rate may actually be decreasing when viewed over a long timescale. Greenland was about as warm or warmer in the 1930s and 1940s, and many of its glaciers may have been smaller then as well. Antarctic ice, which is twenty times more expansive, has been growing at about 1% per decade.

Erratic and unpredictable glacier changes are influenced by a host of different natural factors. A study reported in the May 2012 issue of *Science* examined 200 of them across the Greenland continent between 2000-2010 using radar data collected from synthetic aperture satellites. It found that their individual flow rates varied both in location and time.

Glaciers with growth rates that were accelerating during a few years, decelerated in others. Some accelerating glaciers were in proximity to others that were decelerating. Their individual behaviors were thought to be influenced by a variety of factors, including: fjord, glacier, and bed geometry; local climate; and small-scale ocean water flow and terminus sea ice conditions. Overall, melting speed-ups were much lower than IPCC models projected.[78 79]

---

[77] Braithwaite, R.J. 2002, *Glacier Mass Balance: The First 50 Years of International Monitoring*, Progress in Physical Geography 26: 76-95
[78] Michael Asher, *So Much for Flooded Cities; Greenland Ice Loss Not*

Larry Bell

Sure. But even the *Wall Street Journal* reported "Polar Ice Melt Is Accelerating: Shrinking in Greenland, Antarctica Has Sent Ocean Levels Higher, Study Says". This ain't just your typical run-of-the-mill, left of almost everywhere, *New York Times* or *Washington Post* global warming alarmist stuff. So it must be true…right?

To be more specific, the article discusses a study published in the journal *Science* that states that higher temperatures over the past two decades have contributed to a nearly half-inch rise in global sea levels since 1992, attributing about 30% of that increase to melting of polar ice sheets. The study estimates that roughly half of that 0.43 inch rise was caused by thermal expansion of the oceans (as water warms, it becomes less dense and expands), some from runoff from melting glaciers, and the rest from melting of Greenland and Antarctic ice sheets.

And by the way, it also points out that Greenland has been a bigger contributor to all of that because Northern Hemisphere ocean currents are warmer (yup, it's called the "North Atlantic Oscillation"…a natural cycle that shifts about every 60-70 years), while "Antarctica is so cold that even if warming occurs it won't melt" at the rate seen in Greenland (according to the study co-author).

The study also admitted that it's a tricky question whether or not the overall accelerated melting of polar ice sheets can be linked to man-made climate change influences; that current climate-change models predict that some parts of the Antarctic ice sheets will grow, while others will melt; that Antarctica is not losing ice as rapidly as suggested by many recent studies; and that "The signals suggest there is no immediate threat" from rising sea

---

Increasing, Daily Tech, July 4, 2008
[79] Ian Howat, Ian R. Joughin, and Ted A. Scambos, *Rapid Changes in Ice Discharge From Greenland Glaciers*, Science, February 8, 2007

levels.[80] [81]

In fact, another recent study posted in *Science*, concluded that polar ice sheet melting has been massively overestimated. That analysis is based upon new methods that filter out "noise" from "Gravity Recovery and Climate Experiment" (GRACE) satellite data. As researcher Frederik Simons of Princeton explains, "Our technique learns enough about the noise to effectively recover the signal, and at much finer spatial scales than was possible before."

Simons and his colleague Christopher Haig directed particular attention to the Greenland ice sheet, noting that the Antarctic ice cap is actually getting bigger. While they found that Greenland's ice loss did consistently increase between 2003 and 2010, the change was very patchy from region to region. In addition, the enhanced detail of where and how much ice melted allowed them to estimate that the annual loss acceleration was much lower than previous research suggested, roughly increasing by 8 billion tons annually. Previous estimates were as high as 30 billion tons more per year.

Such rates of Greenland ice loss were barely larger than the margin of error in their readings, making it difficult to discern any difference between a supposed loss curve on a graph from a straight line. At the current rate, it will cause sea levels to rise about 2.4 inches over the next century. And, according to the authors:

> At current melt rates, the Greenland ice sheet would take about 13,000 years to melt completely, which would result in a global sea level rise of more than 21 feet (6.5 meters).

---

[80] *A Reconciled Estimate of Ice-Sheet Balance*, Science

[81] *Grim Picture of Polar Ice-sheet Loss*, Olive Hefferman, November 29, 2012, Nature/News

The good news is that we are scheduled for the next Ice Age long before then. It should give Al Gore at least some comfort knowing that.[82][83]

And Those Dangerously Melting Himalayan Glaciers?

Remember the big crisis after the IPCC warned us that the Himalayan glaciers might disappear altogether by the year 2035, leading to flooding of rivers followed by imminent drought and starvation for billions of people? Well, okay, it later turned out that they made this all up, but never-the-less, those glaciers must be melting pretty fast, or they wouldn't have worried their Nobel Peace Prize-awarded minds and have scared us all to death about it…would they?

Darn! Maybe we all might have gotten a lot more sleep had we known about recent results of another GRACE satellite study published in the journal *Nature* that shows that Himalayan glaciers are hardly melting at all. According to John Wahr at the University of Colorado-Boulder, glaciers and ice caps in places other than Greenland and Antarctica lost about 30% less ice than had previously been estimated.[84][85]

---

[82] *New Science Upsets Calculations on Sea Level Rise, Climate Change Ice Sheet Melt Massively Overestimated, Satellites Show*, Lewis Page, November 28, 2012, The Register

[83] *Princeton Geoscientists Report Greenland Ice Sheet Melting Rate is Increasing*, Bob Yirka, November 20, 2012

[84] *Recent Contributions of Glaciers and Ice Caps to Sea Level Rise*, Thomas Jacob, John Wahr, W. Tad Pfeffer and Sean Swenson, February 23, 2012, Nature

[85] *New Sat Data Shows Himalayan Glaciers Hardly Melting at all: Results 'really were a surprise', say Climate Profs*, Lewis Page, February 9, 2012, The Register

Wahr expressed surprise at these findings, saying:

> *One possible explanation is that previous estimates were based on measurements taken primarily from some of the lower, more accessible glaciers in Asia and were extrapolated to infer behavior of higher glaciers. But unlike the lower glaciers, many of the high glaciers would still be too cold to lose mass even in the presence of atmospheric warming.*

Not only that, it appears that glaciers in the high Asian mountain ranges...the Himalayas, the Pamir and the Tien Shan...were even much better off (presuming the mountains really care). Previous estimates suggesting annual ice depletion rates as high as 50 billion tons were exaggerated by a plus-or-minus 20 billion ton error.

Leveling the Sea Level Discussion

Few (climate alarm skeptics included) are likely to dispute that the world's mean temperatures have been rising at a pretty constant rate of about one degree Fahrenheit (0.6°C) over the past 100 years. Many will also agree that this trend is quite likely to continue, albeit with both warmer and cooler fluctuations, for many hundreds of years into the future. And yes, we can expect seas to slowly continue to rise as well.

Incidentally, neither this general warming trend or sea level rise began with fossil-burning during the Industrial Revolution...nor have they changed in any detectable way due to human influences.

Let's also recognize that sea levels, like climate cycles, really do change and sometimes, but not always, are connected. And are they higher now than in the past? You betcha! In fact they are a lot higher.

Back around 18,000 years ago during the deepest part of the

last Ice Age, a period known as the "Wisconsin", sea levels were about 400 feet lower than now. That was because lots of the water was bound up in land ice. Then about 15,000 years ago huge ice sheets covering North America and Eurasian land masses began to rapidly melt, causing sea levels to rise at the rate of roughly 16 feet per century up until the beginning of the "Holocene Optimum".

Between about 8,000 to 5,000 years ago the rate of rise declined precipitously, and has been relatively stable ever since.

Well…pretty stable. During the Little Ice Age between about 1300-1850, the coldest period of the past 10,000 years, sea levels fell again as ice accumulated in Greenland, Antarctica, Europe and worldwide glaciers.

How do we know about past sea level changes? Since satellite measurements have only become available recently, we must depend upon other indicators.

Proxy records, organic and ocean sediment data present good regional pictures going back many centuries, and tidal gauges provide readings dating back about a century. In addition, we can also get some picture of both temperature and sea level changes over past millennia by looking at melting shrinkage rates of the West Antarctic Ice Sheet. This can be determined by noting how much its "grounding line", the points where it makes contact with the underlying land mass, has receded. Unlike floating sea ice which doesn't influence sea level when it melts, the West Antarctic Ice Sheet is part of the land mass.

See also Appendix 2-4: *Man the Lifeboats! Oceans Rising at Alarming Rate! (Or Maybe not)*, Larry Bell, Forbes Opinions, December 23, 2012

As for evidence that sea levels fluctuate downward as well as up, consider that during a period lasting from the Bronze Age through the Minoan Warm Period and continuing through the Roman Empire, the ancient city of Ephesus had been an important port and commercial hub. Now that city in modern-day Turkey is four miles away from the Mediterranean.

The old Roman port Ostia Antica located where the Tiber River once emptied into the Tyrrhenian Sea is now two miles up-river. A painting by Raphael depicts a sea level high enough to accommodate warships during the Battle of Ostia in 849.

When William the Conqueror defeated King Harold II at the Battle of Hastings in 1066, he landed at an old Roman fort on a small harbor island on England's south coast. That location, now known as Pevensey Castle, is presently a mile from the coast.

And although the UN's IPCC, based upon its highly theoretical climate models, have predicted an increase in the rate of global average sea level rise during the 20th century, that rate has actually been rather stable, with no significant rise over the past 50 years. The rates in the 1920-1945 period were likely to have been just as large as today's.[86 87 88 89 90]

Sea levels have been steadily rising in fits and starts at the rate of about 4-8 inches per century ever since the Little Ice Age ended about 160 years ago. And since they fluctuate over multi-century and multi-decanal periods, any meaningful trends must take long-term perspectives into account. On this basis, there has

[86] Church, J.A., White, N.J., Coleman, R., Lambeck, K. and Mitrovica, J.X., 2004, *Estimates of the Regional Distribution of Sea Level Rise Over the 1950-2000 Period*, Journal of Climate 17: 2609-2625
[87] Cazenave, A. and Nerem, R.S., 2004, *Present-Day Sea Level Change: Observations and Causes*, Reviews of Geophysics, 42: 10.1029/2003RG000139
[88] Douglas, B.C., 1991, *Global Sea Level Rise*, Journal of Geophysical Research, 96: 6981-6992
[89] Douglas, B,C., 1992, *Global Sea Level Acceleration*, Journal of Geophysical Research,97: 12,6999-12,706
[90] Carton, J.A., Giese, B.S. and Grodsky, S.A., 2005, *Sea Level Rise and the Warming of the Oceans in Simple Ocean Data Assimilation (SODA) Ocean Reanalysis*, Journal of Geophysical Research 110: 10.1029/2004JC002817

been no acceleration. For example, while even the 2007 IPCC's summary report concluded that "no long-term acceleration of sea level has been identified using 20th-century data alone", their latest AR5 report (only six years later) somehow disagrees.

Gauging Realities

It's often difficult to determine whether and how much sea levels rise from time to time versus how much the coast lines we measure them from are falling. It's also a big problem determining how much and in what ways temperature changes influence sea levels. Making all of this even more complicated, varying regional geological and climate conditions must also be factored in.

As Dr. Nils-Axel Mörner, former chair of the Paleogeophysics and Geodynamics department at Stockholm University in Sweden points out, tide gauging gives different answers depending upon where they are taken in the world and requires expertise in geology to properly interpret it. Dr. Mörner is just such an expert, having studied effects of sea levels globally over four decades.

A combination of temperature and other non-temperature-related regional factors complicate measurements of sea level trends and consequences. Very long-term natural changes raise the seabed, and subsidence caused by human activities lowers coastal land elevations relative to the sea level.

On a global scale, there are two major temperature influences: land-based snow accumulation and melting (particularly Antarctica and Greenland); and expansion of sea water volume as ocean temperature rises.

For example, the melting of glacial ice cover from northern continents over several millennia has caused land surfaces in some regions to rebound...a process called "isostatic adjustment". This is like what a mattress does when you get out of bed, only a whole lot slower. At the same time, many tidal measurement stations have been sinking due to coastal subsidence caused by compaction of sediments due ground water removal.

Based upon an extensive analysis of the subject Dr. Mörner concludes that "...prior to 5000-6000 years before present, all sea level curves are dominated by a general rise in sea level in true glacial eustatic [correlated with water volume and ocean basin size] response to melting of continental ice caps", but that "...sea level records are now dominated by irregular distribution of water masses over the globe...primarily driven by variations in ocean current intensity and in the atmospheric circulation system and maybe even in some deformation of the gravitational potential surface."

Dr. Mörner then goes on to observe that, "With respect to the last 150 years, the mean eustatic rise in sea level for the period 1850-1930 was [on] the order of 1.0-1.1 mm/year," but "...after 1930-40, this rise seems to have stopped" at least up to the mid-60s. Then after that, with the advent of the TOPEX/Poseidon satellite mission the records show: 1) stability between 1993-1996; 2) a high amplitude rise and fall in 1997-1998 during a large El Nino period; and 3) an irregular record between 1998-2000 with no clear tendency.

Dr. Mörner is highly critical of IPCC sea level analyses which have heavily depended upon data taken from only six tide gauges in Hong Kong, an area of known subsidence. Having been an expert reviewer for two of IPCC's previous reports, he also observes that none of the 2007 authors were sea level specialists.

Traditional sea level measurements using coastal tidal gauges present significant limitations and uncertainties. There are only about two dozen stations throughout the world with records dating back to the early 1900s.

And while satellites have an inherent advantage over tidal stations in direct sea level elevation measurements, those observations date back only to 1993. This is a very short time to draw trend conclusions. To make matters worse, the tidal and satellite measurements don't reliably match up.

Much UN publicity has focused upon a supposed rising sea level threat to the Maldives, a nation of about 1,200 small islands in the Indian Ocean that was predicted to be doomed to disappear in 50 years...or at most, 100. Yet Mörner notes that the sea level rapidly dropped there in about 1970, and hasn't risen since. Had there been any rise at all, a tree that existed there at sea level since the 1950s would have been swept away.

The IPCC's 2007 *COP19* climate conference public education materials entitled *Do People Have an Influence on Climate Change?* claims that "during the last hundred years...the sea level increased for the first time since the last Ice Age (over 20cm since 1870, and the pace of the increase is getting faster), glaciers melt and the snow cap of the Northern Hemisphere decreases."

Curiously, IPCC's 2013 report states, "It is likely that GMSL [Global Mean Sea Level] rose between 1920 and 1950 at a rate comparable to that observed between 1993 and 2010."

While highly theoretical IPCC climate models have predicted an increase in the rate of global average sea level rise during the 20th century, that rate has actually been rather stable, with no significant rise over the past 50 years. And while a satellite-altimeter study published in 2005 revealed an increase, this was most likely believed to be a non-permanent feature of the global ocean's transient thermal behavior that isn't attributable to melting of land-based glacial ice. The global sea level rise rates in the 1920-1945 period were probably just as large as today's.

Based upon records taken from 57 US tide gauges with data base lengths between 60-156 years (with a mean span of 82 years), there had not been any acceleration on the rate of rise along US shorelines over a period of time when alarmists had claimed that the planet had warmed at a rate and to a level "unprecedented" over the past one to two millennia. Instead, those records detected a slight deceleration of -0.0014 mm/year/year.

Twenty-five tide gauge records that contained data for the period 1930-2010 revealed a ten times larger deceleration of minus 0.0130 mm/year/year. Similar decelerations were found in analyses of worldwide gauge records, leading researchers to ask why slight worldwide temperature increases haven't produced an acceleration of global sea level over the past 100 years; and why that level has possibly decelerated for at least the past 80 years?

Yet nothing here is intended to suggest that rising sea levels, however gradual, shouldn't constitute a concern for public preparedness. North Carolina's coastal land slopes, for example, are so gentle in many areas that a one-foot rise can flood areas two miles inland. New Orleans was built below sea level from the beginning. In such cases, not concentrating population centers and vulnerable facilities in hurricane-prone coastal flood plains is a certainly a very prudent idea.[91][92]

At the same time, regarding those rising tides of terror, please someone tell Al Gore that maybe it's time to relax.

When co-anchor Katie Couric asked him on the May 24, 2006 *Today* show "What do you see happening in 15 to 20 years if nothing changes?" "Even Manhattan would be in deep water", he replied. "Yes, in fact the World Trade Center Memorial site would be underwater."

Then, much more recently, when hurricane Sandy blew in, he knew exactly who to blame.

Us, of course...due to our coal-fired-electricity-powered laptop computers, fossil-fueled-climate-ravaging SUVs, and a host of other tide-raising influences.

So what does all of this really mean?

It means that those who spread nonsense that more fossil fuel

---

[91] *The State's Sea Level Retreat*, Orrin H. Pilkey, February 23, 2012, News and Observer

[92] *Commentary: Sea Level Reports Need Clarifying*, Bob Emory, February 26, 2012, Sun Journal

regulations and wind power subsidies are urgently needed to prevent catastrophic coastal flooding are already in way over their heads.

Chapter Nine: Is Your SUV Killing Ocean Coral Reefs?

A MARCH 2012 article appearing in *National Geographic* breathlessly headlined that "Ocean Acidification Rate May be Unprecedented, Study Says." It asserts that "The world's oceans may be turning acidic faster today from human carbon emissions than they did during four major extinctions in the last 300 million years." This distressing announcement projects that a global increase in atmospheric CO_2 concentrations will dramatically increase ocean acidity (lowering the pH), making it increasingly difficult for calcifying organisms such as corals to produce calcium carbonate skeletons. This, they warn, can lead to diminished populations—or even their extinction—by the end of this century.[93]

Carbon Culprit Strikes Again

The alarm got off to a strong beginning thanks to a 2003 study published in *Nature* based upon theoretical geochemical, ocean circulation and climate models, along with a hypothesis that post-

---

[93] *Ocean Acidification Rate May Be Unprecedented*, Study Says, Ira Block, March 1, 2012, National Geographic

21st century fossil fuel burning might cause atmospheric CO_2 concentrations to approach 2,000 parts per million by about the year 2300. The following year, the Pew Center on Global Climate Change, a group advocating immediate action on global warming, gave that wrecking ball of anxiety a big shove. They issued a report predicting that the air's increased CO_2 content could produce a decline in ocean surface pH which might dramatically decrease coral calcification rates, leading to "a slow-down or reversal of reef-building and the potential loss of reef structures." [94 95 96]

A 2009 article published in *Science* reported that coral calcification rates for one species on the Great Barrier Reef had declined 14% between 1990 and 2005, which was claimed to be "unprecedented" in at least the past 400 years. Not surprisingly, this ominous pronouncement was not wasted by the mainstream media. The headline of a *BBC News* feature proclaimed "coral reef growth is slowest ever." A *Sky News* headline read "Barrier Reef Coral Growth Will Stop". *ABC News* topped them all, quoting the research paper's senior author as saying "coral growth could hit zero by 2050."

Not mentioned was the fact that the major reef-builders of today, the scleractanian corals, have been around for some 200 million years. During most of that time the planet's atmospheric CO_2 concentrations were two to seven times higher than now, and temperatures were as much as 10-15 degrees Celsius warmer. And between 1572 and 1605 when atmospheric CO_2

---

[94] *Climate Change Reconsidered*, Science and Environmental Policy Project 2009 Interim Report, Heartland Institute

[95] Caldeira, K. et al., 2003, *Anthropogenic Carbon and Ocean pH*, Nature

[96] Buddemeie, R.W., et al., *Coral Reefs & Global Climate Change: Potential Contributions of Climate Change to Stresses on Coral Reef Ecosystems*, The Pew Center on Global Climate Change

concentrations were about 100ppm lower, the calcification rate of that coral was nearly one-quarter less. (Incidentally, numerous studies show that that no simple linkage has been established between higher temperatures and coral bleaching.)[97]

Recent ocean acidification alarm has been based largely upon IPCC "business-as-usual" projections that CO_2 levels might about double from about 400ppm now, to about 800ppm by the end of this century. According to one scenario, this would cause the ocean surface pH to drop from a preindustrial mean value of 8.2 to about 7.8. A 2010 study headed by Carles Pelejero concluded that the oceans have already acidified by an average of 0.1 pH unit since the preindustrial era.

Other researchers believe that the pH decline expected by Pelejero and his colleagues is much greater, perhaps double, that which is most likely to occur. Pieter Tans predicted in a paper published in a special December 2009 issue of *Oceanography* that atmospheric CO_2 concentrations will peak well before 2100 at only about 500ppm, and then drop back to approximately current levels by 2500.[98]

Ocean surface pH levels often change back and forth over short intervals. In the North Atlantic near Bermuda they ranged from between 8.18 to about 8.03 during different times between 1984 and 2007. Even larger seasonal variations have occurred in some ocean basins. Over only about a decade in the mid-twentieth century, the pH at Arlington Reef in Australia's Great Barrier Reef ranged from 8.25 to a low of about 7.71. At Molokai Reef in Hawaii, seawater pH ranged from a high of 8.29 to a low of 7.79 over just two days in July 2001.[99]

---

[97] De'ath, G. et al., 2009, *Declining Coral Calcification on the Great Barrier Reef*, Science, 323

[98] Tans, P. 2009, *An Accounting of the Observed Increase in Oceanic and Atmospheric CO_2, and the Outlook for the Future*, Oceanography 22

[99] Pelejero, C, et al., 2010, *Paleo-perspectives on Ocean Acidification*,

Larry Bell

Good News for Some Carbonated Calcium Critters

Research reveals that high atmospheric CO_2 concentrations can have important ocean pH enhancement influences as well, stimulating increased carbon-fixing photosynthesis rates by algae. Studies of a wave-exposed boulder reef on the northern New Zealand coast and a sheltered shallow-water area in a Danish fjord showed that daytime pH levels were significantly higher in spring, summer and autumn than in winter, often reaching values of 9 or more during peak summer growth periods vs. 8 or less in winter. At one site, the pH values ranged from 9 during daylight hours, down to below 8 at night.[100]

And while some calcifying species have reportedly exhibited reduced calcification and growth rates under higher pH conditions in laboratory experiments, others actually have higher rates under lower levels. Japanese researchers studied how various concentrations of CO_2 infused directly into filtered seawater to lower the pH influenced the growth and survival of two coral species during a spawning event. They found that larva survival rates of one (*Acropora digitifera*) weren't significantly influenced by different pH conditions, while that of the other (*Acropora tenuis*) was about 18.5% greater in the lowest pH (highest CO_2)...equivalent to an atmospheric CO_2 concentration of 2,000ppm.

This is two and one-half times higher than IPCC has predicted by the year 2100!

While at the end of the study the A. *tenuis* polyp size was reduced by about 14% in the lowest pH treatment, this was more than compensated by a greater survivorship rate. Accordingly, the

---

Trends in Ecology and Evolution 25
[100] Middleboe, A.L, and Hansen, P.J., 2007, *High pH in Shallow-Water Macroalgal Habitats*, Marine Ecology Progress Series 338

researchers concluded that "coral larvae may be able to tolerate ambient pH decreases of at least 0.7 pH units." [101]

Coral calcification, which requires energy, sometimes occurs at the competing expense of protein growth which is also important for coral development. A study conducted by Israeli, French and UK. researchers investigated this circumstance involving two colonies of massive *Porites* corals (which form large multi-century-old colonies that calcify relatively slowly) as well as four colonies of branching *Stylophora pistillata* coral (a short-lived species which deposits its skeleton rather quickly). Fragments of the corals were grown in tanks at pH levels corresponding with CO_2 concentrations upwards to about five and ten times today's atmospheric level.

Although fragments of each species survived and added new skeleton calcium carbonate at a somewhat smaller rate under reduced pH conditions than those grown under normal pH concentrations, the researchers also reported that "tissue biomass (measured by protein concentration), was found to be higher in both species under increased CO_2." Since tissue thickness and protein concentrations are considered to be good indicators of colony health, the authors predicted that the coral species "will be able to acclimate to a high CO_2 ocean even if changes in seawater pH are faster and more dramatic than expected." [102]

But what about ocean acidification influences upon a larger variety of calcifying species than just coral? A study conducted by Woods Hole Oceanographic Institution researchers reared 18 different shell-forming varieties in experimental seawaters with pH levels equal to average atmospheric CO_2 values up to three and one-half times the highest IPCC projection. Their results

---

[101] Suwa, R., et al, 2010, *Effects of Acidified Seawater on Early Stages of Scleractinian Corals (genus Acropora)*, Fisheries Science

[102] Krief, S., et al.2010, *Physiological and isotopic responses of scleratinian corals to ocean acidification*, Geochimica et Cosmochimica Acta, 74

showed that ten of the species had least comparative shell growth in the highest simulated atmospheric CO_2 equivalent, while four had net increases relative to the control under intermediate CO_2 levels (606ppm and 903ppm).

However in three species (crabs, lobsters, and shrimps), net calcification was greatest under the highest level of CO_2 (2,856ppm). One species, the blue mussel, exhibited no response to elevated CO_2. These findings led the team to conclude that "the impact of elevated atmospheric CO_2 on marine calcification is more varied than previously thought"...with responses ranging from negative, to neutral, to positive. In other words, extremely high atmospheric CO_2 concentrations would produce winners and losers.[103]

How typical are such conclusions as those referenced? A review reported in the 2010 issue of *Estuarine, Coastal and Shelf Science* surveyed 372 experimental ocean acidification research studies that investigated 44 different marine species. Of these, 47 of the studies found "no significant response", and only a minority demonstrated "significant responses to acidification." In many cases where significant responses were reported, they typically occurred at atmospheric concentrations exceeding 2,000ppm. At an upper limiting range of 731-759ppm, just below the level predicted by IPCC for the end of the 21st century, calcification rate reductions of only 25% were observed for those minority species.

As the Woods Hole researchers point out, that news may be even better because "most experiments assessed organisms in isolation, rather than [within] whole communities." For example, they note that "sea-grass photosynthesis rates may increase by 50 percent with increased CO_2, which may deplete the CO_2 pool, maintaining an elevated pH that may protect associated calcifying

[103] Ries, J.B., et al., *Marine Calcifiers Exhibit Mixed Responses to CO_2-induced Ocean Acidification*, Geology, 37

organisms from impacts of ocean acidification."

Finally, they conclude that the world's marine biota are "more resistant to ocean acidification than suggested by pessimistic predictions identifying ocean acidification as a major threat to marine biodiversity"...and that this phenomenon "may not be the widespread problem conjured into the 21st century" by the IPCC. Here, once again, IPCC prognostications have hyped doom and gloom without addressing a full range of phenomena and uncertainties that responsible science demands.[104]

And Yes, Human Impacts Do Matter

Recognizing that coral reefs provide vital habitats for wide varieties of life, all influences upon their health should be taken seriously. But Gary Sharp, a marine biologist and scientific director at the Center for Climate/Ocean Resources Study in Salinas, California points out that "We need to look closely at what is most likely to affect the reefs and what is not."

Dr. Sharp observes that conjectures that global warming will kill reefs are based upon highly unlikely predictions that sea temperatures may increase about 3.6°F over the next hundred years, and also that rising CO_2 levels are making oceans dangerously more acidic.

At the same time, there are many human-caused as well as natural environmental stressors that influence reef health and survival. Included are: bacterial infections; releases of man-made toxins (such as industrial wastes, herbicides, pesticides) and poisonous metabolic wastes from marine organisms; excess nutrients from farmland runoffs and sedimentation from coastal and riverbed erosion; damage from monsoons and dredging; and fishing pressure, tourism and boat anchorage.

---

[104] Hendriks, I.E., et al., 2010, *Vulnerability of Marine Biodiversity to Ocean Acidification: A Meta-Analysis*, Estuarine, Coastal and Shelf Science

The good news is that you can still drive your SUV with a pretty clear conscience. Just be careful where you park your skiff in those shallow coral waters.[105]

---

[105] Gary Sharp, *Coral Bleaching: What (or Who) Dunnit?*, Technology Commerce Society Daily, April 26, 2006

Chapter Ten: Biodiversity Bombshell Blows Up on Alarmists

PROMOTING GLOBAL WARMING alarmism has become an effective manipulation tactic to advance a variety of special-interest agendas that often have little to do with the environmental goals and social benefits espoused by responsible individuals and organizations. In 2006, for example, the Institute for Public Policy Research, a think tank that actually supports CO_2 cuts, provided an analysis of circumstances surrounding global warming debates that were occurring in the UK:

> Climate change is most commonly constructed through alarmist repertoire as awesome, terrible, immense and beyond human control...It is typified by inflated or extreme lexicon, incorporating an urgent tone and cinematic codes. It employs [a] quasi-religious register of death and doom, and it uses language of acceleration and irreversibility.

The IPPR concluded that "alarmism might even become secretly thrilling"—effectively a form of what they referred to as "climate porn".

The media has readily bought into this too-good-to-waste

crisis messaging. It has thrilled the public with powerful statements and graphic images that leave lasting impressions of global warming devastation, present and future represented to be based upon science. It's awesome to imagine the Ice Age glacier-threatened Manhattan they once warned about now depicted under water; heartbreaking to see polar bears and penguins becoming extinct, and terrifying to think of horrible diseases that are spreading—all because of us.

As E. Calvin Beisner observed in a June 4, 2010 *Washington Times* article "Move Over, Global Warming—Biodiversity is the Next Central Organizing Principle of Human Civilization", that climate catastrophe alarmism follows a familiar tactic. It is typically based upon computer model projections and hypotheses, not supported by observable empirical evidence. A central premise holds that our modern industrial, agricultural, capitalist society promotes population growth and consumption that is not "natural".[106]

Following this theme, saving Earth from catastrophic man-made climate change has served as the central United Nations rallying mantra over more than two decades. Now, as public warming fears continue to cool, it is cranking up the thermostat on biodiversity alarm. Addressing attendees at the UN's October 2010 biodiversity conference in Nagoya, Japan, Environment Minister Ryo Matsumoto's opening remarks were reminiscent of proclamations broadcast at all of the annual global warming summits. "We are now close to a 'tipping point'—that is, we are about to reach that threshold beyond which biodiversity loss will become irreversible, and may cross that threshold in the next 10 years if we do not make proactive efforts for conserving biodiversity."[107]

---

[106] The Washington Times, E.Calvin Beisner, June 4, 2010

[107] *Ten Years to Solve Nature Crisis, UN Meeting Hears*, Richard Black, October 18, 2010 BBC News: Science and Environment

Climate Crisis Spokesbears

Lots of this fright campaign targets impressionable young minds and sensitive big hearts with messages of fear and guilt. For example, a January 20, 2008 *CBS Evening News* global warming special hosted by Scott Pelly reported that polar bears "may be headed for extinction", noting that researchers are finding them thinner and weaker, with less time to stock up on fat reserves because ice sheets are melting too fast.

CBS reporter Daniel Sieberg, in an August 14, 2007 segment of the *Evening News*, explained that "Less ice also means the polar bears spend more time in the water, sometimes for so long that they drown". Never mind that, as even Polar Bears International points out, swimming up to a hundred miles out to sea is no big deal for these intrepid animals.

ABC's Sam Champion told *Good Morning America* audiences on February 8, 2008 that a 2-degree increase in global temperatures would make "polar bears struggle to survive." On November 6, 2007, *NBC's Today Show* co-host Matt Lauer said the bears "are facing an epic struggle for survival." Reporter Kerry Sanders warned "If the Arctic ice continues to melt, in the next 100 years, the US Wildlife Service says the only place you'll find a polar bear on Earth will be at the zoo." [108]

An April 2006 *Time* magazine cover featured a bear seemingly "stranded" on melting ice. The Defenders of Wildlife website stated, "Loss of sea ice leads to higher energy requirements to locate prey and a shortage of food. This causes higher mortality rates among cubs and reduction in size among first-year adult males."

On March 28, 2008, Paul Milikin, a *Natural Geographic* photographer, stated on ABC's *Good Morning America*, "I realize

---

[108] Nathan Burchfiel, *Polar Bear Scene Could Maul Energy Production*, May 7, 2008

what I need to do is try and tell these stories through *Natural Geographic* magazine by using animals, such as polar bears, to say that if we lose sea ice in the Arctic, and projections are to lose sea ice in the next twenty to fifty years, we ultimately are going to lose polar bears as well."

Milikin went on to acknowledge how the photograph featured on *Time's* cover in 2006, the seemingly "distressed" polar bear came about:

> *It was just a moment where I was not thinking clearly. I was ten feet away, lying on my belly, and this bear is shaking water. And I was just...he took a lunge at me basically, but as [he] lunged up and was coming down on me, the ice broke away. And my first thought was, 'I know I have a shot', so I was really excited that this shot would help tell the story that I want to tell about melting ice.*" [109]

How dire was this survival threat? On a September 9, 2007 *Good Morning America* broadcast, Kate Snow called polar bears "the newest victims of global warming." The same segment featured Dr. Steven Amstrup, a US Geological Service scientist who stated that "bears could be absent from almost all their range by the middle of this century."

It might be interesting to note that only five years earlier, a 2002 study by that same USGS had reported that the "[polar bear] populations may now be near historic highs."

A 2006 US Department of Interior news release stated that it would consider further polar bear protection programs, and the agency acknowledged that "Alaska populations have not experienced a statistically significant decline, but Fish and Wildlife

---

[109] Scott Whitlock, *ABC's Sam Champion Hypes Global Warming for Eight Minutes*, NewsBusters.org, March 28, 2008

Service biologists are concerned that they may face such decline in the future." FWS then requested nine administrative reports from government agencies to bolster its case for listing the bears as a threatened species.[110]

All those reports were based upon climate models that shared common assumptions about sea ice levels during the 21st century—namely, that the area of the Arctic covered by sea ice in summer would decline by more than two-thirds, causing seal populations to decline. No ice, no seals, no bears; case closed.

And they were wrong on all accounts.

Again, let's take another look at those climate-threatened polar bears...the ones adrift and stranded on melting ice caused by our coal-fired power plants and oil-fueled SUVs. A federal investigation into those claims has seriously questioned that. It seems a 2006 paper in the journal *Polar Biology* indicating that "drowning-related deaths of polar bears may increase in the future if the observed trend of regression of pack ice and/or longer open-water periods continues" lacked any real evidence.

That conclusion, which may have been largely responsible for getting polar bears listed as a threatened species was based upon a sighting of four bear carcasses from an aircraft at an altitude of 1,500 feet over the Beaufort Sea that likely died during a storm. Biologist Charles Monnett, the lead scientist on the paper, has returned to work after being placed on administrative leave over the matter.

Quite obviously, his own livelihood isn't threatened. He presently manages $50 million in studies at the Interior Department's Bureau of Ocean Energy Management, Regulation and Enforcement. What have been endangered, however, are any

---

[110] *Interior Secretary Kemthorne Announces Proposal to List Polar Bears Under Endangered Species Act*, U.S. Department of Interior, Press Release, December 26, 2006

near-term prospects for Arctic drilling.[111]

In 2011, the World Wildlife Fund's climate blog headlined that "Polar Bear Population in Canada's Western Hudson Bay Unlikely to Survive Climate Disruption." But it seems that since then this subpopulation, previously believed to be among the most threatened subpopulations due to global warming, has made a miraculous recovery.

According to aerial surveys released by the Government of Nunavut in April 2012, their numbers are at least 66% higher than expected. This region which straddles Nunavut and Manitoba is critical because it's considered to be a bellwether for how well polar bears are faring elsewhere in the Arctic.

And to top off that happy news, recent space satellite images reveal that 36 colonies of Antarctic emperor penguins are twice larger than researchers previously thought. In fact four additional colonies that scientists hadn't known about were discovered as well.[112][113]

Economic Impacts of Fishy Science

Then there's the matter of those minnow-size delta smelt that were determined to be endangered by agricultural and urban fresh water diversions from California's San Joaquin and Sacramento Rivers into their briny eastern marsh habitats. Based upon their listing by the California Fish and Game Commission as an endangered species, water channeled to the Central Valley was cut by up to 90%. This led to 40% unemployment in the San

---

[111] *Global Warming Link to Drowned Polar Bears Melts Under Searing Fed Probe*, Audrey Hudson, August 8, 2011, Human Events, Posted by James Pat Guerrero, August 13, 2011, Wordpress

[112] *Polar Bear Population Growth Confounds Libs*, April 6, 2012

[113] *Emperor Penguins are Teeming in Antarctica*, Robert Lee Hotz, April 14, 2012, Wall Street Journal

Joaquin Valley, turning that major food basket into an empty dust bowl.

There are some big questions about the basis for that decision as well. Two scientists, Frederick V. Feyer of the Bureau of Reclamation and Jennifer M. Norris of the Fish and Wildlife Service, were called to task for presenting misleading court testimony. Regarding Dr. Norris, US District Judge Oliver Wanger commented:

> *I find her testimony to be that of a zealot. I'm not overstating the case, I'm not being histrionic, I'm not being dramatic. I've never seen anything like it. And I've seen a few witnesses testify.*

He went on to say:

> *Does the court reasonably rely on this kind of analysis? What the court uses as the term to describe it is opportunistic. It is an answer searching for a question. It is an ends/means equation where the end justified the means no matter how you got there. Whether you use statistics, whether you use anything that is objective or not.* [114]

A spotted owl protection effort killed logging and created ghost towns throughout the Northwest, only to later discover that the kindness campaign made little difference. Government studies ultimately revealed that those spotted owls weren't logging casualties at all. Instead, they were being victimized by their cousins, the barred owls, who crowded them out of habitats and attacked them. So the government then came up with a $200

---

[114] *More Interior Scientists Are Taking Heat*, Felicity Barranger, September 21, 2011, New York Times

million "barred owl removal plan" to literally shoot the interlopers, a subspecies of the same Mexican owl clan. This has come to be a very familiar solution...namely, for government to choose losing favorites and kill strong competitors.[115][116]

Paramecia in Peril!

Now, just when growing public immunity to feverish global warming hype is relieving hallucinatory sweats, another climatic crisis looms nigh. While some of it is still attributable to "climate change" along, with other human-caused dilemmas, there is a big difference.

Of course this one is even much worse. I'm referring here to mass extinctions of species we don't yet even know about...not to mention even some that we do. This constitutes nothing less than a planetary biodiversity crisis!

The terrifying tale got a big boost a decade ago when Harvard ant biologist Dr. Edward O. Wilson estimated that 50,000 species are going extinct. Yet when environmental activist Tim Keating of Rainforest Relief was asked if he could name any of them he replied, "No we can't, because we don't know what those species are. But most of the species we're talking about in those estimates are things like insects and even microorganisms." It seems they primarily inhabited the computer hard drive that generated his theoretical model.

Regarding Wilson's predictions, UK scientist and professor emeritus of Biogeography at the University of London Philip Stott commented, "The Earth has gone through many periods of major extinctions, some much larger than even being contemplated

---

[115] Tom DeWeese, *Stupid Human Tricks: The Sad Case of the Spotted Owl*, July 2, 2007
[116] *Blasting Some Owls to Save Others?*, Nancy Grace, CBS News, April 27, 2007 CBS.com Stories

today." He went on to say "...the idea that we can keep all species that now exist would be anti-evolutionary, anti-nature and anti the very nature of the Earth in which we live." [117]

Adding fuel to the fire of extinction frenzy is a March 4, 2011 paper published in the journal *Nature* proclaiming "World's Sixth Mass Extinction May be Underway: Study". It states that:

> Over the past 540 million years, five mega-wipeouts of species have occurred through naturally-induced events. But the new threat is man-made, inflicted by habitation loss, over-hunting, over-fishing, the spread of germs and viruses and introduced species, and by climate change caused by fossil-fuel greenhouse gases. [118]

Greenpeace co-founder and ecologist Dr. Patrick Moore believes that the paper is seriously flawed and should never have made it through the peer-review process. In an interview posted on *Climate Depot* he observed that:

> Since species extinction became a broad social concern, coinciding with the extinction of the passenger pigeon, we have done a pretty good job of preventing species extinctions.

He also believes that "The authors [of the journal *Nature*] study greatly underestimate the rate new species can evolve, especially when existing species are under stress," noting: "The polar bear evolved during the glaciations previous to the last one, just 150,000 years ago."

Still, the alarmists' message is clear. If we don't begin to curb carbon-fueled capitalism and transfer governance and unfair

---

[117] 2000 Documentary, *Amazon Rainforest: Clear-Cutting the Myths*
[118] Climate Depot/ Yahoo News

wealth to the UN in penance for our prosperity, many thousands of as-of-yet undetermined insects, microbes and other species are most surely doomed!

Can you live with that guilt?

See also Appendix 2-5: *Biodiversity Bombshell: Polar Bears And Penguins Prospering, But Pity Those Paramecia!*, Larry Bell, Forbes Opinions, April 24, 2012

And What About Those "Bad" Bugs?

Then again, might a little mass extinction of some pesky little buggers be okay…like the ones Al Gore's warned about in his pestilent *An Inconvenient Truth* movie and book? Doesn't human-caused global warming cause lots of really nasty tropical diseases to spread?

Well, maybe not. At least Dr. Paul Reiter, a medical entomologist and professor at the Pasteur Institute in Paris doesn't think so. Dr. Ritter was also a contributory author of the IPCC 2001 report who resigned because he regarded the activities to be driven by political agenda rather than science. In fact he later threatened to sue the IPCC if they didn't remove his name from the report he didn't wish to be associated with.[119]

Professor Reiter's career has been devoted primarily to studying such mosquito-borne diseases as malaria, dengue, yellow fever, and West Nile virus, among others. He takes special issue with any notion that global warming is spreading such illnesses by extending the carriers to formally colder locales where they didn't previously exist.

In reference to statements in *An Inconvenient Truth* that the American cities of Nairobi and Harare were founded above the mosquito line to avoid malaria, and that now the mosquitoes are

---

[119] Paul Reiter: *Global Warming Won't Spread Malaria*, EIR Science & Environment, April 6, 2007: 52-57

moving to those higher altitudes, Dr. Reiter comments:

> *Gore is completely wrong here—malaria has been documented at an altitude of 8,200 feet—Nairobi and Harare are at altitudes of about 4,920 feet. The new altitudes of malaria are lower than those recorded 100 years ago. None of the 30 so-called new diseases Gore references are attributable to global warming. None.*

Although few people seem to realize it, malaria was once rampant throughout cold parts of Europe, the US, and Canada, extending into the 20th century. It was one of the major causes of troop morbidity during the Russian/Finnish War of the 1940s, and an earlier massive epidemic in the 1920s went up through Siberia and into Archangel on the White Sea near the Arctic Circle. Still, many continue to regard malaria and dengue as top climate change concerns—far more dangerous than sea level rise.

Dr. Reiter submitted written testimony to the British House of Lords, 2005. His testimony included the following critique of the chapter written by Working Group II—much of which was devoted to mosquito-borne diseases, principally malaria—for the IPCC's Second Assessment report:

> *The scientific literature on mosquito-borne diseases is voluminous, yet the text references in the chapter were restricted to a handful of articles, many of them relatively obscure, and nearly all suggesting increasing prevalence of disease in a warmer climate. The paucity of information was hardly surprising. Not one of the lead authors had ever written a research paper on the subject! Moreover, two of the authors, both physicians, had spent their entire careers as environmental activists. One of these activists has published 'professional' articles as an 'expert' on 32 subjects, ranging from mercury poisoning*

to landmines, globalization to allergies, and West Nile virus to AIDS.

Species Come and Go Without Our Help

There can be no doubt we humans through a wide variety of our activities influence lives of other critters, and not always for the better. For example, we displace and change natural habitats and ecosystems with urban development and agriculture, intentionally and accidentally introduce non-native animal, insect and plant species, and design domesticated mutations we deem preferable to our interests at evolutionary warp speed...think dogs, cats and cattle.

It would be a highly short-sighted, however, to imagine that mass species changes and extinctions are a recent phenomenon. The great Permian extinction about 250 million years ago witnessed the disappearance of about 95% of ocean dwellers and 70% of land dwellers including many species of fish, corals, reptiles, insects and plants. That was just one of five mass extinctions we know about, along with more than twenty smaller ones. No longer do endless varieties of dinosaurs roam our planet. They disappeared, as did mammoths and woolly rhinos.

Sadly, many species quietly disappear all the time, victims of a changing climate and physical environment, volcanism, meteor impacts, aggressive predators, and constant competition for food and living space. Species, like recurrent Ice Ages, come and go, often in a natural cause and effect sequence. It's quite likely in fact that the vast majority of all species that once lived on Earth have passed through the exit door to oblivion. Those least able to adapt lead the way, and there are no lifetime guarantees for any.

We humans are part of that naturally changing world. We have survived thus far through adaption to changes, often severe, through application of our brains, hands, creative tool-making, and abilities to pass on experiences and ideas through inventions of language.

Try and wish as we might, we cannot preserve every dying species through misguided attempts to reverse human progress. To do so would be at our own peril.

Section 3

Colorblind "Green Energy" Madness

BEWARE OF MARKETING terms such as "clean", "renewable" and "sustainable". While those words may seem very nice, they have routinely been co-opted and redefined through misleading "green" messaging campaigns.

Regarding "clean" energy, let's agree that no sane person wants dirty, polluted air, land or water. Nope, not even conservatives.

At the same time, let's also not confuse carbon dioxide, the plant food that all carbon-based life depends upon, with "air pollution". The simple fact that the Supreme Court gave EPA permission to regulate it as a polluting "endangerment" under its Clean Air Act does not make this true.

That 2009 ruling decreed that atmospheric concentrations of six greenhouse gases (including CO_2) "threaten the public health and welfare of current and future generations". How? Because they pollute global temperatures—warm apparently being considered to be "bad", and conversely, frigid obviously being "good".

The junk science premise for this was even refuted at the time by EPA's own in-house Internal Study on Climate conclusions. Authored by my friend Alan Carlin, then a senior research analyst at EPA's National Center for Environmental Economics, the report stated:

> ...given the downward trend in temperatures since 1998 (which some think will continue until at least 2030), there is no particular reason to rush into decisions based upon a scientific hypothesis that does not appear to explain most of the available data.

So far that internal assessment is very much on track. We have witnessed flat mean global temperatures for going on 18 years now. Even the UN's alarmist Intergovernmental Panel on Climate Change (IPCC) has finally been forced to admit that their climate models have grossly exaggerated climate warming influences of CO_2.

It was political science, not climate science that prompted EPA's decision. As then-presidential candidate Barack Obama promised in 2008 while pushing a "green energy" agenda:

> Under my plan of a [carbon] cap-and-trade system, electricity rates would necessarily skyrocket. Coal-powered plants, you know, natural gas, you name it, whatever the plants were, whatever the industry was, they would have to retrofit their operations. That will cost money. They will pass that money on to consumers.

Let there be no doubt that he has kept his promise to make energy expensive for American citizens and job-producing industries through regulatory strategies where government picks winners and losers.

Most often the losers win. Remember Solyndra?

As for any "alternatives" with sufficient capacities to seriously supplant world fossil and nuclear dependence, don't count on this any time soon. They don't exist, and they certainly aren't cheap. The general public is largely unaware how expensive that "free" and "sustainable" energy really is. Much of the real cost necessary to sustain it is passed on to taxpayers and consumers through invisible subsidies and preferential purchase mandates.

According to 2013 US Energy Information Agency figures, those alleged dastardly climate-killing CO_2-belching coal plants produced 39% of all US electricity, while wind accounted for 4.13% and solar a whopping 0.23%. Nuclear provided 19%, and hydropower (the renewable many "environmentalists" also love to hate) generated 7%.

But then again, there is no alternative source that environmental lobbies universally love. Utility-scale solar power systems that produce electricity using thermal collectors or photovoltaic cells draw eco-attacks for taking up too much desert land, thus displacing certain animal and reptile species. Those which use photovoltaic collectors are also challenged because they are manufactured using highly toxic heavy metals, explosive gases and carcinogenic solvents that present end-of-life disposal hazards.

The best wind power sites are typically along mountain ridges and coastal areas, locations also prized for scenic views and over-flown by bird and bat species which become turbine blade casualties. Few people want to live anywhere near them due to noise and other psychological—even physical—health impacts.

But what about biofuels produced from organic plants? Aren't they supposed to offer a "green" way to reduce dependency upon those brown fossil fuels for heating and transportation? Well...not really.

Corn ethanol yields less energy than is required to grow and produce it, competes for land with food production, and releases copious CO_2 emissions (if you really care). And that much-touted cellulosic ethanol from plant waste remains a long way off from commercial reality.

It is essential to our national energy future that the voting public becomes much better informed about advantages and disadvantages of all alternatives, and that crony capitalist lobbies not be allowed to hijack beneficial free market competition. "Green energy" has become a meaningless term, where capacities and benefits have been grossly exaggerated, and where industry sustainability depends upon endless preferential government-rigged charity.

Chapter Eleven: No Free Power Lunches

OKAY, CALL ME a spoilsport, but as much as we all might wish otherwise, there really is no free energy lunch. Not so long as much or more energy needs to be put in to get it than it yields, it removes more valuable land and other resources from beneficial uses than it is worth, it isn't reliably available in sufficient quantities when and where needed, it promises highly debatable environmental advantages, it costs consumers lots more than "nonrenewable" alternatives, and its sustainability depends upon endless government mandates and subsidies.

The time is past due to challenge those who would have us believe that "next generation" biofuels, wind farms and solar power meet necessary criteria, whereas fossil and nuclear energy sources by definition do not. Such arguments often fail to acknowledge that capacities to serve demands, economies of production and unsubsidized competitive free market pricing are fundamental sustainability requisites.

Might evolutionary technology developments change this picture? Yes, maybe in certain cases and to some degree that could happen. But don't count on any fossil and nuclear alternatives making much of a difference any time soon, regardless how much taxpayer green we throw at them.

119

See also Appendix 3-1: *Green Energy Promotions Rely Upon Colorblind Public*, Larry Bell, Newsmax, September 8, 2014

Since just about everything we do and the equipment needed to support it depends upon a source of energy, wouldn't it be great if someone would invent perpetual motion machines that can generate all we want without consuming any resources or producing pollution? Okay, some of you are doubtless saying, "Yes, and they already exist. There are wind turbines and solar power systems that can do that if we build enough of them."

Sorry…the solution just isn't that easy.

First of all, without reversing progress back to the Stone Age (and even then, remember those smoky caves), we couldn't create adequate numbers of either or both to accommodate modern power demands regardless how much conservation we practiced. One constraint is suitable land area. There simply aren't enough appropriate wind and solar site locations to make that happen.

Another limitation is power supply unreliability. Even gargantuan installations covering thousands of acres generate small only amounts of expensive and intermittent power.

Wind Power's Overblown Prospects

Wind simply doesn't afford the benefits marketing blowhards promise. It isn't an abundant, reliable power source; doesn't appreciably reduce fossil dependence or CO_2 emissions; isn't free, or even cheap; doesn't produce net job gains; nor does it cool brows of feverish environmental critics.

Many green energy advocates grossly exaggerate the capacity of wind power to make a significant impact on US electrical needs. A common ploy is to conflate maximum projected total capacities, typically presented in megawatts (MW), with actual predicted kilowatt hours (kWh) that are determined by annual average wind conditions at a particular site. Wind is intermittent, and velocities constantly change. It often isn't available when needed most—such as during hot summer days when demands for air-conditioning are highest.

Output volatility due to wind's intermittency varies greatly according to location and time of year, typically ranging from 0% to about 50%. Texas, one of the most promising wind energy states, averages about 16.8% of installed capacity, yet the Electric Reliability Council of Texas assigns a value of 10% due to unpredictability. Only about 20% of that capacity is generally available during peak demand periods (about 5:00pm), while average generation during off-peak time averages about 40% of capacity.[120][121]

Electricity must be instantaneously available day and night to meet "base load" requirements. When peak loads exceed supplies bad things quickly happen. Electrical frequencies and voltages drop as power line currents increase, necessitating automatic or manual interruption of loads (blackouts) to protect grids.

But unlike such workhorse power generators as coal-fired and nuclear plants designed to constantly run at peak load capacities, wind (and solar) power requires incorporation of "spinning reserve" backup systems to provide continuity. These are typically gas-fired turbines, much like those used for jet aircraft engines that are connected to generators. That's where it gets particularly expensive.

Wind power must be integrated as part of a larger, balanced, grid network. When that wind generation component increases, the temperatures of fossil-fueled boilers must be dropped to maintain demand-supply equality. This involves wasteful shedding of heat for cooling—then more wasting to add heat back into the

[120] Industrial Wind Action Group, *Transmission Issues Associated with Renewable Energy in Texas*, March 28, 2005
[121] ERCOTT: Tudor, Pickering, Holt & Co., Energy Investment and Banking, *Texas Wind Generation Report*, 2009

system without accomplishing any additional work. And since the spinning reserves don't stop consuming fuel when wind generation is occurring, claims of energy savings or CO_2 emission reductions are largely mythological.

Another major limitation of individual wind farms is that they don't produce power on massive scales needed in large cities and industrial areas where necessary space is at a premium and land is expensive. The most ideal locations are typically remote from areas where demands are highest, requiring large investments for power transmission lines and land right-of-way use.

Wind turbines are also very expensive to build and maintain. The National Renewable Energy Laboratory reports that "despite reasonable adherence to those accepted design practices, wind turbines have yet to achieve their design life of 20 years, with most requiring significant repair before the intended life is reached". Those in offshore locations are even more costly to install, and fare much worse from corrosion damage.[122]

Will the construction and maintenance of wind power produce the many thousands of "high-quality green jobs" touted by the industry? Not according to a report from Spain released by researchers at King Juan Carlos University. It concluded that every "green job" created by the wind industry killed off 4.2 jobs elsewhere in the Spanish economy through missed opportunities to put that money towards more useful and productive ends.

While research director Gabriel Calzada Alvarez didn't fundamentally object to wind power, he did find that when a government artificially props up the industry with subsidies, higher electrical costs (31%) and tax hikes (5%), along with government debt follow. Each of those jobs was estimated to cost

---

[122] National Renewable Energy Laboratory, *Improving Wind Turbine Gearbox Reliability*, W. Musial, S. Butterfield and B. McNiff, May 7-10, 2007

$800,000 per year to create, and 90% of those were temporary. A few months after the study was released, researchers at the Danish Center for Politiske Studier reached similar conclusions based upon their country's experience: "It is fair to assess that no wind energy would exist if it had to compete on market terms".[123]

Just how environmentally friendly is that "green" wind energy? Depends a lot on who you ask and where they live. While some national environmental organizations such as Greenpeace and the Sierra Club have become staunch wind power advocates in their war against fossils, others who live in proposed wind farm locations have launched strong legal opposition.[124]

But what about risks to our economy and the well-being of ratepayers and taxpayers who must cover wind power costs?

More than half of all revenues for companies that install and operate the systems come from federal, state and local tax benefits. Some state programs also legislate mandatory renewable portfolio standards that require electric utility companies to purchase designated amounts of energy from wind, solar and bio-fuel providers, typically at premium costs that are passed on to customers.

So long as industry survival depends upon those preferential government-imposed benefits, two things are clear. Wind is certainly not a competitive free market source of energy, or a charity we can continue to afford.

See also Appendix 3-2: *Wind's Overblown Prospects*, Larry Bell, Forbes Opinions, March 2011

---

[123] Investor's Business Daily, *The Big Wind-Power Cover-Up*, March 12, 2010
[124] Los Angeles Times, *First U.S. Offshore Wind Energy Project Faces Lawsuit*, Tribune Washington Bureau, Kim Geiger, June 26, 2010

Solar Power's Cloudy Future

Solar power, like wind, is a natural, free source of energy— provided that public subsidies and customers of high-priced electricity cover the costs. In the US the main federal subsidy currently pays for 30% of the cost for a residential system. Then when other subsidies are added, as much as 75% of the cost can be covered. The US Energy Information Administration estimates that electricity produced by solar is presently three times more expensive than power produced by natural gas.[125]

Also, like wind, it is severely constrained by geography and is intermittent. And even under the best geographical sky and climate conditions there is still a recurring problem. It's called "night".

While the solar power industry has been growing, thanks largely to taxpayer photovoltaic panel and installation subsidies and a glut of cheap Chinese government-subsidized system imports, it is doing so from a very small base. All of the panels now installed across the nation produce only about as much electricity as a single coal-fired plant (about 0.01% of US electricity). In fact Kentucky's Cardinal coal mine alone produces about 75% of the BTU energy generated by all US solar panels and wind turbines. Yet the Sierra Club boasts that it has "unplugged" more than 100 US coal fired plants.[126]

A photovoltaic system with enough arrays to power Los Angeles would require blanketing thousands of square miles of desert habitat. Or just replacing the electrical output of the Palo Verde Nuclear Power Station near Phoenix with the type of photovoltaic technology used at Nevada's Nellis Air Force Base

---

[125] *Solar Power Showing Greater Mainstream Potential*, October 23, 2011, Chicago Sun-Times

[126] *Sustainability: Some Free Marker Reflections*, Marlo Lewis, February 11, 2011, MasterResource

would require solar arrays covering an area ten times larger than Washington, D.C., and at costs 15 times higher per kWh. By contrast, we could produce an estimated 670 billion gallons of oil from frozen tundra equal to one 20th of that area if drilling in the Arctic National Wildlife Refuge was permitted.[127][128]

Green Power Gridlock

Managing the uninterrupted transfer of electrical power from myriad sources wherever and whenever it is needed is a hugely complicated challenge. It's one thing when the principal supply sources use gas, heat or hydraulically-driven turbines which provide constant, unfluctuating outputs that can be adjusted and counted upon independent of weather or season.

Circumstances become increasingly complex as more and more intermittent sources are added to the power supply mix. Difficulties arise as segments of the grid become overloaded or underserved by the renewables, requiring turbines which balance the grid to be constantly throttled down. This reduces turbine operating efficiencies.

Utility grid operators are sometimes forced to dump wind energy produced on blustery days when regional power systems don't have room for it. As Long Beach Mayor Bob Foster, a board member of the California grid system management commented, "We are getting to the point where we will have to pay people not to produce power."

Adding more and more wind and solar to the power grid is creating an increasingly uncertain and dangerous juggling act. Infrastructure development and management involves integration and control of a vast patchwork of power lines and monitoring

---

[127] *Renewables are Unsustainable*, Paul Driessen, CFACT
[128] *Solar Power is Beginning to Go Mainstream*, Jonathan Fahey, Associated Press/ USA Today

devices connecting industrial-scale fossil, nuclear, hydro, wind and solar electricity-generating plants. All of this juggling is subject to hazards of tripping on a tangle of antiquated and changing legal market rules, operational formulas and business models.

Adding greatly to the challenge, uncertainty regarding continuing availability of large government subsidies and other industry perks required to make renewables "competitive" in the marketplace presents another big challenge. This makes long-term system infrastructure planning virtually impossible.

At least one green energy developer recognizes that stimulus subsidy programs have a record of doing more harm than good. Patrick Jenevein, CEO of the Dallas-based Tang Energy Group, posted a *Wall Street Journal* article arguing that:

> *After the 2009 subsidy became available, wind farms were increasingly built in less-windy locations...The average wind-power project built in 2011 was located in an area with wind conditions 16% worse than those of the average...Meanwhile, wind-power prices have increased to an average $54 per megawatt-hour, compared with $37 in 2005.*[129]

Subsidies obviously influence markets. Writing in the *LA Times*, Evan Halper quotes Neil Fromer, the executive director of Caltech's Resnic Sustainability Institute, observing that:

> *One of the biggest challenges is you can't create a market for the resources without solving the demands of moving electricity from one physical place to another. But you can't solve that problem until you understand what the*

---

[129] *Wind Power Subsidies? No Thanks*, Patrick Lenevein, April 1, 2013, Wall Street Journal

Larry Bell

market structure will look like.[130]

Trieu Mai, a senior analyst at the National Renewable Energy Laboratory laments, "The grid was not built for renewables." Accordingly, as Halper notes, "Planners are struggling to plot where and when to deploy solar panels, wind turbines and hydrogen fuel cells without knowing whether regulators will approve the transmission lines to support them."

Nevertheless, the problems can only get worse as California leads other states in a rush to bring more and more wind and solar power onto grids that weren't planned to accommodate it. A report by a group of Caltech scholars projects that the necessary upgrades to make a green future work will be "one of the greatest technological challenges industrialized societies have undertaken". They project this can be expected to cost about $1 trillion nationwide by 2030.

And what is likely to occur if that additional taxpayer and ratepayer subsidized cost burden isn't covered? Jan Smuthy-Jones, executive director of Sacramento's Independent Producers Assn. which represents owners of renewable and gas power plants, presents an ugly scenario. He warns that current proposals to move California to as much as 80 percent renewable energy within the next two decades are bumping up against prospects of another San Diego-type blackout which occurred in 2011.

On that blistering hot day streetlights went dark, flights were grounded, pumping station failures caused sewage to flow onto beaches, and people were trapped in office elevators and Sea World rides. All of those consequences, and more, were caused by an employee error at a power substation near Yuma, Arizona.

Will grid limitations put a damper on successes of climate alarmists and other anti-fossil activists to push costly non-

[130] *Power Struggle: Green Energy Versus a Grid that's not Ready*, Evan Halper, December 2, 2013, LA Times

alternative energy technologies into ever-more risk-prone grids? This remains to be seen. However as our older nuclear plants are decommissioned and new EPA regulations shutter coal-fired plants, one thing is certain. States like California that continue to increase renewable requirements are likely to resemble Europe in more ways than even they wish to emulate.

See also Appendix 3-3: *Green Power Gridlock: Why Renewable Energy Is No Alternative*, Larry Bell, Forbes Opinions, December 10, 2013

Lessons from Lemmings: the EU's Breezy Folly

Lemming powers of observation aren't highly regarded. Wouldn't you think witnessing fellow critters plunge en masse over cliff edges would offer cause for some among them to reconsider the perilous path ahead? But then some humans, including government leaders, are similarly lacking in foresight. This deficiency is evident in the EU's wind and solar energy stampede which has speeded their arrival at the brink of economic collapse. Before America races any farther down that same road, let's pause to heed lessons from their costly experiences. And we might learn much from our own as well.[131][132]

Existing policy in Germany already forces households to fork out for the second highest power costs in Europe—often as much as 30 percent above the levels seen in other European countries. Only the Danes pay more, and residential electricity costs in both countries are roughly 300 percent higher than in the US.

Approximately 7.8 percent of Germany's electricity comes from wind, 4.5 percent from solar, 7 percent from biomass, and 4 percent from hydro. The government plans to increase the proportion from renewables to 35 percent by 2020, and 80

---

[131] Global Wind Energy Council
[132] *The Wind Experience*, Institute for Energy Research

percent by 2050. Since hydro and biomass won't grow, most of that expansion must come from wind and solar.

Denmark meanwhile, which produces between 20 to 30 percent of its electricity from wind and solar, hopes to produce half from those sources by 2020. As Denmark can't use all the electricity it produces at night, it exports about half of its extra supply to Norway and Sweden. But even with those export sales, government wind subsidies have led Danish consumers to pay the highest electricity rates in Europe.

And what about Britain? In 2011, UK wind turbines produced energy at a meager 21 percent of installed capacity (not demand capacity) during good conditions. As in Germany, unreliability in meeting power demands has necessitated importation of nuclear power from France. Also similar to Germany, the government is closing some of its older coal-fired plants—any one of which can produce nearly twice the electricity of Britain's 3,000 wind turbines combined.

Renewable source unreliability is less of a problem when there are reliable backup sources like hydropower, coal and nuclear plants to meet demand. But most of Europe lacks the former, and is intentionally—to its detriment—cutting back on both of the latter.

Facing future winter blackouts, a recent series of measures set to be taken by Britain's National Grid includes asking owners of decommissioned gas, coal and oil power stations to turn them back on as well as compensate offices and factories to shut down power use for four hours per day between November and February so that energy can be diverted to households. As UK Independence Party energy spokesman Roger Helmer MEP commented to *Breitbart London*:

> *UKIP has been warning for years that the UK's energy policies, driven by Brussels, would lead to blackouts and supply shortages...Now, the chickens are coming home to*

roost.

MEP Helmer continued:

> *It is a bitter irony that DECC [the Department of Energy and Climate Change] is planning contracts with commercial companies to use diesel generators to fill the gaps when the wind doesn't blow. It is bizarre that we are paying over the odds for diesel generators when the plan was to cut emissions. The near-certainty of blackouts is not the only problem. We've also forced energy prices sky high, driving businesses and jobs and investment out of the UK, and leaving households and pensioners in fuel poverty.*[133]

Signs of constructive change are far more apparent in Australia. In September, the right-of-center Liberal Party's defeat of the Green Party-backed Labor Party was recognized as a referendum victory for dismantling and consolidating myriad anti-carbon schemes spawned under the previous government. Reining in that sprawling climate machine and eliminating an established energy carbon tax is expected to save the economy more than Au$100m (£57.6m) per week. Australia is seeing sense—there are lessons Europe can learn from its shining example.

And thanks to natural gas, coal and nuclear, the US has excess power-generating capacity, and generally adequate transmission and distribution systems—unlike Europe. Today, just over 42 percent of US electrical power comes from coal, 25 percent from natural gas, and 19 percent from nuclear. Only about 3.4 percent comes from wind, and about 0.11 percent from solar.

---

[133] *Britain Faces Winter of Blackouts as Firms Are Asked to Ration Electricity,* Donna Rachel Edmunds, September 3, 2014

Whether renewable energy will be able to offer substantial cost-competitive alternatives—rather than limited niches—for US and international energy remains to be seen. But regardless, we can only hope that America learns from the ruinous green energy policies in Germany and other EU nations before such misguided policies wreak further social and economic damage.

Meanwhile, so long as natural gas drilling is restricted, climate crisis hoax-premised EPA regulations strangle fossil power generation, and nuclear energy expansion is delayed, we are racing hell-bent along the same road to perdition. Let's consider the ultimate lemming peril before joining them in a final, fatal jump.[134][135]

---

[134] *Misguided Energy Policies Have Put Europe on a Path to Economic Decline*, Larry Bell, City AM newspaper (London), October 11, 2013

[135] *Lessons from Lemmings: The E.U.'s Green Power Folly*, Larry Bell, Forbes Opinions, October 3, 2011

Chapter Twelve: Biofuels, Fields of Pipedreams

"PLANT IT AND they will come." Sure, but just when might we begin to expect those benefits; the ones promised with passage of costly federal legislation such as the Renewable Fuels Act (2005) and the Biofuels Security Act (2006)? We're still waiting for some sign of taxpayer and consumer paybacks—or perhaps at least some hopelessly die-hard optimists are.

Ethanol refiners and other advocates typically cite energy independence as a compelling argument for the massive subsidies. Unfortunately, any notion that we can ever fuel our way towards energy security by planting waving fields of grain is seriously misguided. Those who would have us believe otherwise grossly exaggerate potential capacities and ignore unpleasant consequences. Both deceptions are prevalent in marketing hype that tells us what we might really wish to believe, namely that biofuels offer Earth-friendly, sustainable fossil alternatives that can wean us away from our "oil addiction". Unfortunately, this is not the case.

For starters, let's focus upon corn ethanol (more commonly known as grain alcohol), since it is currently the only domestically-produced commercial biofuel. It really isn't renewable at all when you consider that nearly as much fossil fuel-generated energy is required to produce it as it actually yields.

Then consider the land use requirements. An attempt to produce enough ethanol to replace gasoline altogether would require that about 71% of all US farmland be dedicated to energy crops. By way of illustration, let's just think about brewing all of our present US corn production into 180-proof grain alcohol. That would displace, at most, about 14% of the gasoline we presently guzzle.[136]

Cellulosic ethanol produced from switch grass and other low maintenance plant materials wouldn't compete nearly as much for land needed for food crops, but has proven to be far more difficult to process than promoters have suggested. The generous government subsidies we have previously provided haven't succeeded in nudging them anywhere close to commercial viability.

Ethanol Mandate Support May be Running out of Gas

About 35% of all US corn grown is being consumed by the ethanol industry, producing nearly 14 billion gallons of alcohol. Congress originally mandated that ethanol blended into US gasoline will increase to 35 billion gallons per year by 2022. This would require that crop land dedicated for this purpose be expanded from 88 million acres now to about 233 million acres (slightly more than half of our 461 million acres total crop land to meet about 7% of our total automotive fuel needs).

An *Associated Press* article titled "The Secret Environmental Costs of US Ethanol Policy" concluded that "The ethanol era has proven far more damaging to the environment than politicians promised and much worse than government admits today." It

---

[136] *Undoing America's Ethanol Mistake*, Senator Kay Bailey Hutchinson, April 28, 2008

observed that ethanol mandates have spurred farmers to grow corn on relatively unproductive land that would otherwise have remained undeveloped.

AP reported:

> *Five million acres of land set aside for conservation— more than Yellowstone, Everglades and Yosemite National Parks combined—have vanished on Obama's watch. Landowners filled in wetlands. They plowed into pristine prairies, releasing carbon dioxide that had been locked in the soil. Sprayers pumped out billions of pounds of fertilizer, some of which seeped into drinking water, contaminated rivers and worsened the huge dead zone in the Gulf of Mexico where marine life can't survive.*

The AP article went on to opine that:

> *The consequences are so severe that environmentalists and many scientists have now rejected corn-based ethanol as bad environmental policy.* [137]

Since US farmland is scarce and expensive, each additional acre of corn used to produce ethanol is one less that is available for other crops such as soybeans and wheat, which have seen dramatic price increases. This, in turn, produces a ripple effect that raises costs of meat, milk, eggs, and other foods with both national and international consequences. Since US farmers provide about 70% of all global corn exports, even small diversions for ethanol production have produced high inflation levels in America and shortages abroad. [138, 139]

---

[137] Boston Herald/AP, *The Secret Environmental Costs of U.S. Ethanol Policy*
[138] *Ethanol Isn't Worth Costlier Corn Flakes and Tortillas*, Michael Economides, May 17, 2011, Forbes.com

Larry Bell

All That Grows is Not Green

And after all, unlike fossil fuels, wasn't biofuel supposed to be "green"? Actually, it's a lot browner than advertised. Even though ethanol fuel may produce marginally less CO_2 emissions than gasoline does (in case you really care about that), this doesn't account for the CO_2 emissions released during corn planting, fertilizing, harvesting and distilling, which on balance, pretty much nullifies any difference. Burning ethanol also releases large quantities of nitrogen oxide (smog) that causes respiratory disease.[140]

In fact there is growing evidence that biofuels may actually release more CO_2 emissions than conventional petroleum-based petroleum does. As reported in the journal *Science*, "Corn-based ethanol...instead of producing a 20% savings, nearly doubles greenhouse emissions over 30 years." This is because big biofuel markets encourage farmers to level forests and convert wilderness areas that serve as CO_2 sinks into farmlands.[141]

Ethanol also competes with people and livestock for water—lots and lots of water. It requires about 4 gallons of water to make one gallon of alcohol fuel. This is in addition to other water that production facilities typically recycle. Many Corn Belt regions where distillers are sited, particularly in the Midwest and the Great Plains, have already begun to experience significant water supply problems.

Agricultural irrigation and cattle feed lots which are located near the plants to take advantage of the co-product distiller's grain

---

[139] *How Biofuels Could Starve the Poor*, Benjamin Senauer, Foreign Affairs May/June 2007

[140] *Ethanol Comes with Environmental Impact, Despite Green Image*, Tom Davies, May 5, 2007, USA Today

[141] *Biofuels May Emit More Greenhouse Gas*, Matt Ball, February 9, 2008 Vector 1 Mediz.com

add to local water demands. Consequently, aquifers in many areas are being depleted faster than they can recharge.[142][143]

There's also a big water pollution problem. Ethanol mandates are prompting more and more corn to be planted on land that is poorly suited for agriculture, causing erosion and pesticide runoff to infiltrate groundwater and aquifer resources. Rather than rotating corn planting with soybeans to replace soil nitrogen, many farmers are planting corn year after year and adding large amounts of nitrogen fertilizer.

On average, about 30 pounds/acre of each 140 pounds/acre of nitrogen fertilizer leaches away and runs off into creeks, lakes and aquifers.

Even more runoff occurs when corn isn't rotated with other crops because the soil develops clumps requiring more tilling that loosens it, resulting in more erosion. Some of that polluted runoff winds up in drinking water, posing special health problems for children and pregnant women.[144]

Ethanol transportation imposes additional difficulties and costs. Unlike oil and natural gas, it can't be moved through existing pipelines because it readily absorbs water and various impurities. Instead, it must be transported by truck or rail, either of which is more expensive.

Then there's an even more fundamental problem with ethanol...namely that it is lousy and expensive fuel. Since the energy density is about one-third less than that of gasoline, more must be burned to produce the same amount of power, translating into reduced gas mileage per gallon.

---

[142] *Water Use by Ethanol Plants*, Dennis Keeney and Mark Muller, October 2006, Institute for Agriculture and Trade Policy
[143] *Ethanol vs. Water: Can Both Win?*, Sea Stachura, September 18, 2006 Minnesota Public Radio
[144] Ibid

A Political Wakeup Call for EPA

Negative political consequences are finally even being recognized by the EPA. Although the agency was required by law to finalize Renewable Fuel Standards which establish ethanol blending requirements for 2014 by November 30, at the time of this writing this has not occurred. The intent was to reduce America's dependence upon foreign energy.

In November 2013 the agency proposed to reduce corn-produced ethanol in 2014 from the 14.4 billion gallons initially required by Congress in the 2007 to 13.01. Overall, EPA proposed to require transportation fuel companies to blend 15.21 billion gallons from all ethanol sources including corn, soybeans and other products into the nation's fuel supply in 2014, down from 16.55 billion gallons in 2013.

As noted by James Taylor of the Heartland Institute: "The reduced ethanol blending requirement reflects various concerns regarding fuel supply and the ethanol mandate. Gasoline demand has been lower than recent EPA forecasts, meaning less ethanol is needed to meet desired blending percentages. Also, advances in cellulosic ethanol and other next-generation ethanol sources are not occurring as quickly as EPA anticipated. Finally, a growing amount of data indicates ethanol is not as environmentally friendly as ethanol advocates claimed." [145] [146]

Panic in the Corn Fields

Ethanol producers found good reason to panic when renewable fuel champion California Democrat Senator Dianne Feinstein

---

[145] *EPA's Ethanol Mandate for 2014 Behind Schedule*, Christopher Doering, Des Moines Register, Monday, June 30, 2014
[146] *EPA Reduces 2014 Ethanol Mandate*, James Taylor, The Heartlander, December 24, 2014

teamed up with Oklahoma corn state Republican Tom Colburn sponsored a bill in December 2013 to get rid of Renewable Fuel Standards. They cited fears that corn-based fuel production mandates will harm livestock producers. Poultry companies are also feeling distress due to rising feedstock prices as corn crops are diverted to ethanol.

And that's not all. RFS mandates requiring refiners to blend higher percentages of ethanol into gasoline to meet EPA's requirements prompt engine fueled equipment manufacturers and consumer groups to warn that automobiles, boats, lawnmowers, and other motorized machines suffer damage to key parts when gasoline blends include more than 10 percent ethanol.

In December 11 testimony to the US Senate Environment & Public Works Committee, even Christopher Gundler who heads EPA's Transportation and Air Quality Office admitted that ethanol mandates are pushing gasoline refineries close to the point where adding more ethanol to the fuel mix is counterproductive. He testified "We're recognizing that the blend wall has been reached."[147]

Why to Care that E15 Ethanol is on the Way to Your Gas Station

In addition to delivering poor efficiency E10 and greater ethanol blends absorb water, causing them to eat up fiberglass boat fuel tanks and rubber fuel line gaskets, clog up valves, and totally destroy small two-cycle engines. (Trust your mechanic on these matters, not what you here from the Renewable Fuels Association trade organization.)

Yet in August of 2013 the US Court of Appeals rejected a

---

[147] *Producers Panic as Ethanol Mandate Loses Support*, James Stafford, Free Republic, January 2, 2014

challenge by automakers and other groups seeking to overturn the EPA's previous approval of E15 automotive fuel containing 50% more ethanol. Brought forth by the Alliance of Automobile Manufacturers, Global Automakers, the Grocery Manufacturers Association and the petroleum industry, the suit charged that the provision would likely cause a "concrete" and "imminent" injury to any automaker, refiner or food processor.

The majority opinion held by Chief Justice David Sentelle and Judge David Tatel, found that the petitioners lacked standing to sue, arguing that refiners and food producers are not injured because EPA is merely giving refiners the "option" to switch from E10 to E15...not a requirement to do so. Yet in practical terms, this is a distinction without any difference, since increased use of ethanol mandated by the Renewable Fuel Standard (RFS) will essentially force them to do so.

So what does this mean to you? Possibly more than you realize. Let's consider a few reasons why.

According to a study prepared by FarmEcon, ethanol E10 already added about $14.5 billion in automotive fuel costs during 2011 due to higher energy costs and negative effect on fuel mileage. This amounted to about 10 cents more for each gallon of US gasoline. Ethanol tax credits (since discontinued), added another 4 cents/gallon.[148][149]

Check to be certain that E15 doesn't nullify your car warranty. Since like all alcohol it is "hygroscopic", it absorbs large amounts of water molecules that combine with petroleum to cause premature rust. It is also a powerful solvent that attacks rubber seals and plastic parts used in engine components, causing them to dissolve, stretch and wear out, or become dry and brittle.

---

[148] *Ethanol Added $14.5 Billion to Consumer Motor Fuel Costs in 2011, Study Finds*, Marlo Lewis, July 19, 2012, GlobalWarming.com
[149] *The RPS, Fuel and Fuel Prices, and the Need for Statutory Flexibility*, Thomas E. Elam, July 16, 2012, FatmEcon LLC

While EPA says tests show that E15 won't harm 2001 and newer vehicles, that claim is disputed. According to findings of the Coordinating Research Council and other organizations, as many as 5 million cars manufactured since 2001 could have engines damaged by running hotter fuel. Two (Toyota and Lexus), have put labels on gas caps warning that their engines are not designed to operate with E15, and that they won't be held responsible for damage caused by higher blend gasoline.

A cautionary note from the National Automobile Dealers Association states that "If drivers mistakenly put E15 in their tanks and their vehicles aren't designed to burn it, they could risk damaging their engines. Car and truck owners may contact their dealership's service department to determine any fuel restrictions." [150] [151] [152]

Ethanol also plays real havoc with boat engines and fiberglass gas tanks. It tends to dissolve and release corrosive matter (gunk) such as resins, varnish and rust which contaminates fuel and travels through marine engines to clog filters, carburetor jets and injectors. Since boats live in a water environment, and ethanol (alcohol) loves to absorb water, use of ethanol above E10 invalidates all marine warranties.

A particularly troublesome issue for boat and fishing enthusiasts is ethanol decomposition of fiberglass gas tanks. The usual fix involves tank replacement, often a costly and time-consuming project, although lining or sealing a tank is sometimes possible for added protection.

The alcohol totally wrecks small engines. Using ethanol blends in two-stroke engines such as mowers and chainsaws results in a low octane mix (lean fuel) which can destroy them.

---

[150] Coordinating Research Council Report
[151] *Auto Industry Adopts New Warning for Gas Cap Labels*
[152] *EPA Approves E15 Fuel Label Despite Engine Risk*, James R. Healy, June 29, 2011, USA Today

Referring to E10 ethanol, Rich Herder, owner of a lawnmower repair business in Westfield, New Jersey, reported *to Popular Mechanics* that "It's the biggest disaster to hit gasoline in my lifetime." He estimates that as much as 75% of his repair work results from use of the blend.[153]

Finally, the E15 mandate—all renewable fuel mandates—exemplify out-of-control government. The E15 mandate, in fact the RFS in general, represents a basic government violation of American free market tenants by favoring one industry over another.

As Circuit Judge Kavanaugh wrote in his dissenting opinion:

> *When an agency illegally regulates an entity's competitor in a way that harms the entity—for example, by loosening regulation on the competitor—we have said that the entity has Article III standing to challenge the allegedly illegal regulation...Here, EPA's E15 waiver loosens a prohibition on gasoline and ethanol producers and thereby harms entities such as the food group that directly compete with gasoline and ethanol producers in the upstream market for purchase of corn.*

Judge Kavanaugh went on to say:

> *Before the E15 mandate, petroleum producers likely could not meet the requirement set by the statutory renewable fuel mandate. Now that EPA has allowed E15 onto the market, producers likely can meet the renewable fuel mandate—but they must produce E15 in order to do so...In the real world, does the petroleum industry have a realistic choice not to use E15 and still meet the*

---

[153] *Can Boutique Fuel Save Small Engines from the Wear and Tear of E10*, Roy Berendsohn, August 12, 2011, Popular Mechanics

statutory renewable mandate? The answer is no, and the intervenor Growth Energy's claim to the contrary seems rooted in fantasy.[154]

It is high time to realize that ethanol alcohol clearly isn't the big renewable and clean energy solution it was brewed up to be.

See also Appendix 3-4: *Ten Reasons to Care That E15 Ethanol is on the Way to Your Gas Station*, Larry Bell, Forbes Opinions, September 23, 2012

---

[154] *Ethanol Mandate Waiver: Decks Stacked Against Petitioners*, Marlo Lewis, September 10, 2012, Global Warming

Chapter Thirteen: Environmentalists Want a "None of the Above"

PRESIDENT OBAMA'S OBSESSION with transitioning from fossil-fueled energy use to so-called "clean renewables" is being thwarted by unlikely adversaries. A 2011 US Chamber of Commerce report titled *Project/No Project* found 140 renewable projects that had stalled, stopped, or been outright killed due to "Not in My Back Yard" (NIMBY) environmental activism and a system that allows limitless challenges by opponents.

The study concluded that it is just as difficult to build a wind farm in the US as it is to build a coal-fired plant, with about 45% of all challenged projects being "renewable energy". This is accomplished by a variety of strategies, including organizing local opposition, changing zoning laws, preventing permits, filing lawsuits, and using other long delay mechanisms, effectively bleeding projects dry of their financing.

The study also confirmed that there were very few "shovel-ready" renewable energy projects that were truly qualified for support under the American Recovery and Reinvestment Act of 2009 (stimulus funding). And according to the US Department of Energy, even if all renewable sources (including hydro) which now provide about 10% of American energy were to grow at three times the pace of all others, they would still make up just

16% of all domestic supplies by 2035.[155] [156]

Absolutely no energy options are immune from environmental challenges. No, it's certainly not just "dirty" coal, oil and natural gas, that are being challenged…or those "hazardous" nuclear plants. Hydroelectric dams are under assault for killing fish, biomass burning produces greenhouse gases just as fossils do, and geothermal power releases toxic ground and water contaminates. Wind turbines slaughter birds and bats, solar power disrupts fragile desert ecosystems.

Wind and solar power also require huge amounts of land and expansive transmission lines to deliver electricity from remote sites. For example, an 85-mile Green Path North Transmission Line planned to carry green power to Los Angeles was cancelled in 2010 due to environmental opposition. As Mike Garland, CEO of Pattern Energy Group, a wind farm developer observed, "We are starting to see all renewable energy projects, no matter how well planned, are being questioned." [157]

Greenies Courting Concerns

In addition to high development and operations costs which are passed on to tax and rate-payers, environmental groups and near-by landowners are fighting so-called "green energy" proposals tooth-and-nail in the courts.

Environmentalists successfully blocked a proposed 500 megawatt wind project on private land in a remote part of Montana near the Canadian border planned by GreenHunter Energy. Plans were shelved after the Montana Wilderness Association and Montana Audubon and Wilderness Society

---

[155] U.S. Chamber of Commerce, Project No Project
[156] *Environmentalists Fight Solar, Wind, Renewable Energy*, August 10, 2012, Investor's Business Daily
[157] Ibid

protested that 400-foot tall turbines would loom over an adjacent wilderness area about 10 miles away. In 2006, GreenHunter announced it would scale down the development to 150 megawatts, then agreed in 2007 to scale back again to only 50 megawatts.

While company officials cited lack of "financial viability" for the project termination, in early 2008 the company representatives had more directly attributed risks of further environmental resistance as the main reason. GreenHunter Chairman Gary Hunter told the Associated Press, "...if you have opposition [to a wind farm] in Valley County, I don't know how you could build one [anywhere]."[158] [159]

In April 2012, two environmental groups, the Portland Audubon Society and Oregon Natural Desert Association, filed suit in the US District Court in Portland to block plans to build an industrial-scale wind energy installation on Oregon's Steens Mountain, along with essential access roads and transmission lines. They charged that the project represented the "antithesis of responsible renewable development" which would threaten golden eagles, sage grouse and big-horn sheep. Approved by then-Interior Secretary Ken Salazar, the $300 million Echanis Wind Project and its 40-60 turbines would cover 10,000 acres of private ranch land with transmission lines extending 44 miles across rolling sagebrush land controlled by the US Bureau of Land Management (BLM).[160]

In October 2012, The Defenders of Wildlife, Center for

[158] Associated Press:
http://www.kulr8.com/news/state/9970866.html;
http://www.macalester.edu/windvisual/valleyinfo.html#sum
[159] Ibid
[160] *Lawsuit Against Wind Energy Project Near Steens Mountain Pits Green Groups Against Green Project*, Richard Cockle, May 2, 2012, The Oregonian

Biological Diversity, and the Sierra Club, filed suit in the Kern County Superior Court in Bakersfield, California to reopen an environmental review of a 100-turbine North Sky River wind project and a smaller adjacent Jawbone project in the mountains north of the Mojave Desert over threats to California condors.

They charged that the installations will also pose hazards to golden eagles, southwestern willow flycatchers and bats. Center for Biological Diversity biologist Ileene Anderson said:

There's plenty of room in the state for both wind projects and the California condor to thrive. But if condors and wind turbines are going to coexist, those turbines need to be sited carefully and measures have to be taken to minimize the risk that condors will be killed. Unfortunately, this project fails to do that.

Plaintiffs filing an injunction claimed they had tried to work with wind developer NextEra to implement project features to minimize the threat to wildlife, but were rebuffed. The groups, backed up by letters of concern from Fish and Wildlife Service (FWS) and the California Department of Fish and Game, asked Kern County to insist on greater environmental protection measures, but Kern's Board of Supervisors approved the project without amendments in September 2011. FWS said that they would not issue take permits for the species, meaning that any wind turbine injuring a condor might face criminal charges.[161]

During 2011, five of nine large solar projects approved the prior year by state and federal regulators in time to qualify for federal loan guarantees became entangled in environmental impact litigation. A lawsuit filed by the Loma Prieta Chapter of the Sierra Club, the Santa Clara Valley Audubon Society, and Save

---

[161] *Court To Consider Injunction In Kern County Wind Case*, Chris Clark, July 27, 2012, Industrial Wind Action Group

Panoche Valley, sought to stop a 399 megawatt, 3,200 acre solar plant in the Panoche Valley, 130 miles southwest of San Francisco. As the plaintiff's attorney told the court, "No one disputes the necessity for solar energy...but it's improper on this site."

The organizations claimed that the project introduced potential risks of hazardous materials and emissions, wildfires and noise which would degrade air quality, prime farmland, cultural resources and endangered species (the blunt-nosed lizard, San Joaquin kit fox and giant kangaroo rat). They charged that many of the inadequacies were caused in a rush to approve the project in order to qualify it for federal stimulus dollars.[162] [163]

In 2012, the Sierra Club, Natural Resources Defense Council (NRDC) and Defenders of Wildlife also filed suit to stop another 663 megawatt 4,600-acre Calico solar plant to be built on 7.2 square miles in the Mohave Desert northeast of Los Angeles. Originally planned to provide 850 megawatts of electricity generated by 30,000 solar dishes standing 40 feet high, the project was scaled back over concern about impacts on desert tortoises. Regardless, the deal ran into more problems with the Sierra Club, which joined with California Unions for Reliable Energy over the developer's hiring of non-union labor. Together, they petitioned the state Supreme Court to block the project on environmental grounds.

After those efforts failed, Tessera sold the project to K Road Power, a New York firm, which decided to switch to solar panels. That still wasn't enough, and the Sierra Club sued the BLM, FWS and the Department of Interior over threats to tortoises and other wildlife. The union group, which had signed a labor agreement

---

[162] *Chapter Joins Suit Against Panoche Valley Solar Plant*, Erin Barrite, July/August, 2011, The Loma Prietan Sierra Club
[163] Ibid

with K Road, didn't participate in the latest litigation.[164]

And what about those solar environmental advantages...like protecting the planet from climate-ravaging carbon dioxide emissions? Well, maybe not after all...at least not according to a letter of protest from three environmental organizations to BLM over new Department of Interior rules to streamline approval for solar energy projects on hundreds of thousands of acres of federal land. The letter complains that "no scientific evidence has been presented to support the claim that these projects reduce greenhouse emissions."

In fact, the letter issued by the Western Lands Project, Basin and Ranch Range, and Solar Done Right indicates that "...the opposite may be true. Recent work at the Center for Conservation Biology at the University of California, Riverside, suggests that soil disturbance from large-scale solar development may disrupt Pleistocene-era caliche deposits that release carbon to the atmosphere when exposed to the elements, thus negat[ing] the solar development C[arbon] gains."

And if this isn't bad news enough, the letter says that the environmental impacts from the solar panels "...are long-term (decades to centuries)" and threaten the habitat of "...endangered species, including the desert tortoise, Mojave fringe-toed lizard, flat-tailed horned lizard, golden eagle and desert bighorn." [165]

Yes, and most of those "green-renewable" projects have lots of other problems as well...namely that they haven't delivered the economic or employment benefits that were advertised. In January 2009, President Obama pledged:

> We will put Americans to work in new jobs that pay well
> and can't be outsourced—jobs building solar panels and

---

[164] *Sierra Club, NRDC Sue Feds To Stop Big California Solar Power Project,* Todd Woody, March 27, 2012, Forbes
[165] Ibid

wind turbines.

But just how well is that green approach working so far? About 20 of those government-backed green energy companies that rushed to get stimulus help ran headlong into financial problems, ranging from layoffs...to losses...to bankruptcies. Yes, in addition to those birds, bats, lizards, kangaroo rats and tortoises, we taxpayers were endangered too.[166][167][168]

See also Appendix 3-5: *Environmental Groups Strongly Endorse 'None of the Above' Energy Plans*, Larry Bell, Forbes Opinions, March 12, 2013

Yes, Let's Have All of the Above We Can Afford

But let's not sacrifice competitive and proven free market principles in the bargain. None of this discussion is intended to suggest that we should plan to rely only upon fossil and nuclear to meet our energy needs, or that conservation strategies aren't at least as important as supply-side initiatives.

It is essential to our national and global future that development and utilization of all alternatives be expanded, but also be realistic about expectations. We must encourage and invest in a broad range of important and promising technologies, including high-density storage batteries, affordable and efficient fuel cells, clean coal and coal-petroleum conversion, safe nuclear and waste storage applications, and a host of transportation,

---

[166] *Panel: Green Jobs Company Endorsed by Obama and Biden Squandered $535 Million in Stimulus Money*, John Rossomando, February 22, 2011, Daily Caller

[167] *More Solar Companies Led By Democratic Donors Received Federal Loan Guarantees*, September 29, 2011, Daily Caller

[168] *Obama's Solar Failures Abound*, March 31, 2012, Investor's Business Daily

heating and power economy innovations to name only a few. In short, we must do more with less wherever possible, and that applies to us all in our personal and business lives.

We should also demand honest information regarding real options, benefits and costs based upon objective science and factual representations. Proposed caps on carbon emissions, whether through crony capitalist cap-and-trade schemes or regressive carbon taxes based upon trumped-up climate hysteria will solve nothing. They will only make energy and other consumer goods more expensive and drive industries overseas along with jobs and tax revenues. The only green to be realized will move from all of our pocketbooks to the bank accounts of carbon hedge fund speculations, politically-connected subsidy recipients and over-promising green energy marketers.

Yes, by all means, let's give all energy options incentives to advance. The best way to accomplish this is to let entrepreneurial innovators and consumers—not government bureaucrats and rent seekers—determine the winners and losers.

Section 4

Regulation Run Amok

FEW SHOULD DISPUTE either that human society doesn't need regulations and means of enforcement to protect precious land, air and water environments from callous abuse and endangerment or that the US Environmental Protection Agency, Bureau of Land Management and other government entities haven't contributed important services to this end. Big problems ensue, however, when monstrous regulatory bureaucracies are allowed to inflict callous abuse and endangerment upon legal and economic structures of the very society they were established to protect. And this is clearly occurring.

EPA, the eight hundred pound, ever power-hungry and growing gorilla of regulation was established by the Nixon administration in 1970 to assemble a patchwork of 15 existing federal environmental agencies and parts of agencies under the control of a single entity. That same year Congress passed the Clean Air Act to set national air standards. A Clean Water Act followed in 1972.

Shortly after opening its doors EPA filed suit against Detroit, Cleveland, Atlanta and other cities for polluting rivers with sewage. That was a good thing to do, resulting in demonstrable improvements.

By the mid-1970s the agency had assumed a broad regulatory stance. Congressional passage of the Federal Insecticide, Fungicide, and Rodenticide Act in 1972, the passage of the Resource Conservation and Recovery Act in 1976, and the reauthorization of the Clean Air Act in 1977 all set in motion the massive regulatory machinery we see in operation today. Air and water programs were placed under one assistant administrator, while another assistant administrator was given responsibility for nearly everything else: pesticides, radiation and solid waste.

Rapid expansion of EPA powers and growth is emblematic of Big Government bureaucracy run amok with the help of some influential friends. In EPA's anti-CO_2 campaign case, there was no better friend than Enron.

Flashing back to the 1990s, Enron was then diversifying its energy business to emphasize natural gas, a fuel that was facing difficult market competition with coal. The company was already heavily invested, and owned the largest natural gas pipeline that existed outside Russia, a colossal interstate network. They badly needed some help in Washington to tip the playing field.

National hype about a global warming crisis advanced by then-Senator Gore's highly publicized 1988 congressional hearings on the subject provided a dream opportunity. Enron hired the event's star witness, NASA Godard Institute for Space Studies director James Hansen, as a consultant, and then began direct discussions with the senator.

Some members of Congress were already aggressively pursuing development of green legislation models with interesting possibilities to advance Enron's purposes. Senators John Heinz (R-PA) and Timothy Wirth (D-CO) had previously cosponsored "Project 88" to provide a pathway for converting environmental issues into business opportunities. Media-fueled alarm about acid rain provided a basis for legislation to create markets for buying and selling excess sulfur dioxide (SO_2) and nitrogen dioxide emission credits, and Project 88 became the Clean Air Act of 1990. Enron had become a big SO_2 market cap-and-trade player.

So Enron and others wondered, why not do the same thing with CO_2? Since natural gas is a lower CO_2 emitter than coal, that development would certainly be a profitability game changer. But there was a problem. Unlike SO_2, CO_2 wasn't a pollutant—at least not then—and the EPA had no authority to regulate it.

After Senator Wirth became undersecretary of state for global affairs in the Clinton-Gore administration, he began working closely with Enron's boss, Kenneth Lay, to lobby Congress to grant EPA the authority to control CO_2. Between 1994 and 1996, the Enron Foundation contributed nearly $1 million to the Nature Conservancy, and together with the Pew Center and the Heinz Foundation, they engaged in an energetic and successful global warming fear campaign that included attacks on dissenters. That was a time when that exact same Heinz Foundation, headed by Teresa Heinz Kerry, awarded a $250,000 scientific award to James Hansen, who then publicly supported her husband in his failed presidential bid.

A September 1, 1998 letter from Enron CEO Lay to President Clinton requested that he "moderate the political aspects" of the climate discussion by appointing a "Blue Ribbon Commission". His intent was clear: to trash disbelievers and cut off debate on the matter. Lay had direct contact with the White House earlier when he reportedly met with President Clinton and Vice President Gore on August 4, 1997, to prepare a US strategy for an upcoming Kyoto climate summit that December.

Kyoto presented the first step toward creating a carbon market that Enron desperately wanted Congress to support. An internal Enron memorandum stated that Kyoto would "do more to promote Enron's business than almost any other regulatory initiative outside the restructuring [of] the energy and natural gas industries in Europe and the United States".

Sadly (for Enron), that was not to be. In a rare spirit of

solidarity, the Senate unanimously passed (95-0) a bipartisan Byrd-Hagel US Senate Resolution (S Res 98) that made it clear that the United States would not be signatory to any agreement that "would result in serious harm to the economy of the United States". Then-President Clinton, no stranger to political pragmatism, got the message and never submitted a necessary US approval request for congressional ratification.

You can bet that Al Gore, who was then gearing up for his own run for Clinton's job in 2000 wasn't one bit happy about that development. A looming global warming crisis was featured as a pitch point in his campaign, with cap-and-trade legislation offered as the road to salvation. Then, as now, he implies that US buy-in to the Kyoto Protocol was Bushwacked by G.W., never seeming to mention the US Senate rebuff. Perhaps he counts on the likelihood that his global audiences are too young to know differently, too old and senile to remember, or just too indifferent to have been paying attention.

See also Appendix 4-1: *EPA and Enron End-Runs of Congress: Lisa Jackson Serves US Industry a Thanksgiving Turkey*, Larry Bell, Forbes Opinions, December 2010

Chapter Fourteen: The EPA's Costly Rampage

AS MY FRIEND, hydrologist Dr. Jay Lehr who drafted many sections of regulation for the EPA during its early days observes, by the 1980s most of the agency's important work had been accomplished. But rather than reduce its largesse, it took on a larger mission. That's when "environmental advocacy groups saw the environment as a way to promote big government and liberal ideas that reduced individual freedom, and threw a monkey wrench in the path of progress and capitalism." [169]

Environmental activists and their captive EPA got their big monkey wrench in the form of an April 2, 2007 Supreme Court case (Massachusetts v. EPA), where the court ruled in a split 5-4 decision that greenhouse gases fit within the definition of "air pollutants" even if science regarding contributions to public health or welfare endangerment are uncertain. The ruling rejected a position taken by the EPA during the Bush-Cheney administration that CO_2 greenhouse gases were not air pollutants.

This development gave the EPA long-awaited authority to

---

[169] Interview with Dr. Jay Lehr, *Defender of Our Industry*, Paradigm and Demographics, March 1, 2011

regulate CO_2 emissions from new cars under its Clean Air Act. The EPA's position was supported in its suit by 12 states (CA, CT, ME, MA, NJ, NM, NY, OR, RI, VT, WA), a number of local governments and environmental organizations. Ten states (AL, ID, KS, MI, NE, OH, SD, TX, UT), along with four motor industry trade associations and two utility company coalitions, opposed granting that authority. Although the ruling did not require EPA to regulate emissions, and was specifically limited to address vehicles, it did pave the way for expansionary developments that followed.

On December 7, 2009, EPA Administrator Lisa Jackson, signed two distinct findings that advanced its reach. One was the "Endangerment Finding" which found that current and projected atmospheric concentrations of six greenhouse gases (including CO_2) "threaten the public health and welfare of current and future generations". As previously mentioned, this was based upon global warming crisis projections put forth by the UN's IPCC which were refuted by EPA's own National Center for Environmental Economics "Internal Study on Climate" report conclusions. A second "Cause or Contribute Finding" found that "combined emissions of these well-mixed GHGs from new motor vehicles and new motor vehicle engines contribute to greenhouse gas pollution which threatens public health and welfare".

On April 1, 2010, EPA finalized a light-duty vehicle rule controlling greenhouse gas emissions, confirming that January 2, 2011, is the earliest date that a 2012 model year vehicle meeting established limits can be sold in the US. Then, on October 25 EPA and the National Highway Traffic Safety Administration (NHTSA) issued a proposed rule to establish the first-ever greenhouse emission and economy standards for heavy-duty trucks which will phase in during model years 2014-2018.

Both organizations argue that improved fuel efficiency growing out of this ruling will save the trucking industry lots of money. Assuming that truckers are pretty smart, and find profit attractive, why should it be necessary to sell the idea using a climate protection ruse? Unlike the light-duty vehicle market, most medium and heavy-duty truck purchasers operate on slim profit margins where fuel costs are a big factor in their economic survival.

In 2010 the EPA also expanded its regulatory reach beyond mobile emissions to include control of GHGs from stationary sources as well. Under rules of its "Prevention and Significant Deterioration" (PSD) program, a central target in the Obama administration's all-out war on fossil fuels is the coal industry. The real casualties of that war will be businesses, jobs and household energy budgets, with few if any public health benefits.

As noted by my friend H. Sterling Burnett, a senior fellow at the National Center for Policy Analysis (NCPA):

> The EPA is in the process of codifying a whole slate of new air quality rules, the sheer number and economic impact of which have not been seen at any time in the EPA's history.

Included are new Mercury and Air Toxics Standards (MATS) for ozone, mercury, and other substances along with greenhouse gases which he predicts "will have an unprecedented negative impact on the US economy." [170] [171]

Since EPA's Boiler MACTS are so strict that not even the best-performing sources can meet them, many companies will have no choice but to shut their doors and ship manufacturing jobs overseas. When combined with other restrictions on coal ash and cooling water that EPA is planning, the impacts will be far-reaching. Credit Suisse estimates that the MATS rule alone could

[170] National Center for Policy Analysis bulletins
[171] *New EPA Air Regs Will Cost Billions of Dollars*, April 11, 2011; *The EPA's New Air Quality Regulations: All Pain, No Gain*, Part One, April 11, 2011

157

lead to closure of nearly 18% of the nation's coal-fired generating capacity and cost the industry $100 billion by 2017. An analysis by the United Mine Workers union concluded that this plan, along with other EPA regulations, could put as many as 250,000 jobs at risk.

In 2011 the American Council for Capital Formation estimated that new EPA regulations already in place would result in 476,000 to 1,400,000 lost jobs by the end of 2014. Management Information Services, Inc. foresaw that up to 2.5 million jobs will be sacrificed, annual household income could decrease by $1,200, and gasoline and residential electricity prices may increase 50% by 2030. The Heritage Foundation has projected that the greenhouse gas regulations will cost nearly $7 trillion (2008 dollars) in economic output by 2029.[172]

Meanwhile, according to an annual "Regulator's Budget" compiled in 2010 by George Washington University and Washington University in St. Louis, the employment of federal government regulators had climbed 13% since Obama took office, while private sector jobs shrank by 5.6%. In fact, if the federal government's regulatory operations were defined as a business, their $54 billion budget would make them one of the 50 the largest in the country...bigger than McDonald's, Ford, Disney and Boeing combined. It's high time we voters issued pink slips to those responsible for mismanaging that bloated enterprise.[173][174]

EPA's End Runs of Congress

The Supreme Court's 2007 ruling and the EPA's endangerment

---

[172] Ibid

[173] *Overruled*, Investor's Business Daily, June 3, 2011, *EPA: Jobs Don't Matter*, Investor's Business Daily, April 19, 2011

[174] *Regulatory Agencies Staffing Up*, Investor's Business Daily, August 16, 2011

finding both predated the ClimateGate scandal which revealed a great variety of gross scientific research survey and reporting improprieties within the IPCC. More recent emails released in connection with these activities add even more reasons to regard IPCC's blatantly agenda-driven processes and conclusions as a very poor basis for government policymaking of any kind.

Nevertheless, shrouded under the ever expanding blanket of the CAA, the endangerment finding is being applied to validate an unprecedented regulatory takeover of carbon-emitting energy and construction industry permitting. This represents a spectacular overreach of authority which end-runs designated responsibilities constitutionally reserved for Congress.

Over the past three decades the Congress not only has never passed any legislation to regulate climate change, but has specifically rejected legislation to give EPA the authority which it now claims. Proposals put forth in 1990 requiring EPA to regulate manufactured substances based upon their "global warming potential" and adopt GHG emission reduction as a national goal failed to pass through the House-Senate Conference Committee on Clean Air Amendments.

An attempt to require the EPA to regulate emissions from automobiles failed in the Senate version before the amendments went to the Conference Committee. In addition, the US Senate preemptively rejected the UN's Kyoto Protocol when it passed the Byrd-Hagel resolution by a vote of 95-0, and Congress enacted a series of funding restrictions that barred EPA and other agencies from implementing the treaty.[175][176]

The failed attempt by former Senator Baucus (D-MT) and other Democrats during the course of 1990 House-Senate Conference Committee on Clean Air Act Amendment proceedings to authorize EPA to regulate CO_2 and other so-called

[175] Marlo Lewis, March 21, 2011, Pajamas Media.Com
[176] *The Senate's Showdown*, March 28, 2011, Wall Street Journal

"greenhouse gases" for climate change purposes would have required the agency to set CO_2 emission standards for motor vehicles. Baucus was also defeated in subsequent proposals to adopt GHG emission reduction as a national goal, or to require EPA to regulate manufactured substances.

Since then, Congress has abandoned its legislative and oversight responsibilities as established by constitutional separation of powers protections in allowing unelected Executive branch agency officials to impose de facto laws through regulatory fiat.

EPA's congressional circumvention actions became more and more extreme at a faster and faster pace under the Obama administration. These include CO_2 emission standards for new power plants which are so strict they will virtually eliminate coal as a fuel option for future electric power generation.

While EPA has punted on standards for existing power plants as well as refineries—standards which will further drive up electricity and gasoline prices, once these regulations are in place, we can expect the agency to proceed under auspices of its Clean Air Act to issue regulations, industry by industry, until virtually every aspect of the American economy is constrained by strict bureaucratic permitting requirements.

How about regulations on wood burning stoves for example? It seems that even wood isn't green or renewable enough anymore. The EPA has recently banned the production and sale of 80 percent of America's current wood-burning stoves, the oldest heating method known to mankind and mainstay of rural homes and many of our nation's poorest residents. The agency's stringent one-size-fits-all rules apply equally to heavily air-polluted cities and far cleaner plus typically colder off-grid wilderness areas such as large regions of Alaska and the American West.

While these most recent regulations aren't altogether new, their impacts will nonetheless be severe. Whereas restrictions had previously banned wood-burning stoves that didn't limit fine airborne particulate emissions to 15 micrograms per cubic meter of air, the change will impose a maximum 12 microgram limit. To put this amount in context, EPA estimates that secondhand tobacco smoke in a closed car can expose a person to 3,000-4,000 micrograms of particulates per cubic meter.[177]

Most wood stoves that warm cabin and home residents from coast-to-coast can't meet that standard. Older stoves that don't cannot be traded in for updated types, but instead must be rendered inoperable, destroyed, or recycled as scrap metal.

The impacts of EPA's ruling will affect many families. According to the US Census Bureau's 2011 survey statistics, 2.4 million American housing units (12 percent of all homes) burned wood as their primary heating fuel, compared with 7 percent that depended upon fuel oil.

Local governments in some states have gone even further even than EPA, not only banning the sale of noncompliant stoves, but even their use as fireplaces. As a result, owners face fines for infractions. Puget Sound, Washington is one such location. Montréal, Canada proposes to eliminate all fireplaces within its city limits.

Only weeks after EPA enacted its new stove rules, attorneys general of seven states sued the agency to crack down on wood-burning water heaters as well. The lawsuit was filed by Connecticut, Maryland, Massachusetts, New York, Oregon, Rhode Island and Vermont, all predominately Democrat states. Claiming that EPA's new regulations didn't go far enough to decrease particle pollution levels, the plaintiffs cited agency estimates that outdoor wood boilers will produce more than 20 percent of wood-burning emissions by 2017. A related suit was filed by the environmental group Earth Justice.

An April 2014 Supreme Court 6-2 decision provides even

---

[177] Editorial: *EPA's Chilling Effect*, Washington Times, January 14, 2014

greater latitude for EPA's runaway legislation-by-regulation zeal. This ruling overturned a D.C. Circuit appellate panel and revived the EPA's 2011 cross-state pollution rule. The EPA's primary target is Texas and other states that have large coal-fired plants.[178]

The latest Supreme Court decision is but one more reminder that we are living in an era of an ever-expanding state of government agency control. As Justice Scalia wrote in minority dissenting opinion:

> *Too many important decisions of the Federal Government are made nowadays by unelected agency officials exercising broad lawmaking authority, rather than by people's representatives in Congress.*

See also Appendix 4-2:*New Congress Must Rein In Runaway EPA*, Larry Bell, Newsmax, June 2, 2014

Unfortunately, if Congress allows these circumstances continue, we ain't seen nothin' yet. Additional EPA restrictions on greenhouse gas emissions, tougher water guidelines and tightening of the ozone standard will continue to drive up pump prices, impose construction bans on local communities, and cripple oil, natural gas and coal production.

EPA's Dirty War on Coal

The EPA has issued a new regulatory assault on coal to prevent a climate crisis that doesn't exist with a mandated solution which remains to be invented based upon claims which are unsubstantiated. They certainly didn't let a Clean Air Act requirement that mandated technologies be "adequately demonstrated" hinder their new performance standards ruling that

---

[178] *EPA's Cross-State Pollution Rule Upheld By Supreme Court*, Dina Cappiello and Sam Hananel, Huffington Post, April 29, 2014

puts a 1,100-pound limit per megawatt hour on carbon emissions from new coal power plants.

Not only is there no scientifically-supportable climate benefit for limiting such emissions, there is no viable commercial-scale technology to achieve that ideological pipedream. Even if it mattered, the most modern coal-fired plants can only reduce CO_2 emissions to 1,800 pounds or more. What's more, they already knew that. A fact-finding working group within EPA's own Science Advisory Board (SAB) informed them that such systems have never been proven outside the laboratory, and that pilot demonstration projects under development are over budget.

SAB issued a memo which raised questions regarding the technical feasibility of "sequestering CO_2," (removing and storing it) noting that "the peer-review of scientific and technical information presented for coal-fueled sources appears to be inadequate." In doing so it contradicted EPA's claims based upon speculative studies and models produced by a Department of Energy research unit that sequestration works.

Although the EPA had assured the SAB panel that those studies had been vetted by "industry experts, academia and government research and regulatory agencies", further investigation revealed that the "peer-review" for some of those studies had been conducted by none other than EPA itself and that others were apparently biased. The SAB working group was unable to obtain a documented description of DOE's peer review process and other information requested from EPA.

EPA officials got around the pesky problem of providing evidence requested by arguing that SAB wasn't supposed to investigate the sequestration science after all. And why? Well according to EPA, they weren't really mandating sequestration or making rules for it. No...they were merely requiring coal plants to meet specific emission limits they set (whether achievable or not). How to accomplish that was a separate issue. That was the industry's problem, not theirs.

Actually, it's our problem too…a very big one. According to the Institute for Energy Research this "regulatory assault" will eliminate 35 gig watts of electrical generating capacity…10% of all US power. As the Competitive Enterprise Institute observes, "If the carbon dioxide emissions standard for power plants proposed by the EPA today is enacted, the United States will have built its final coal-fired power plant."

It was political science, not climate science, which demanded hasty EPA action. As then-presidential candidate Barack Obama promised in 2008 while pushing a CO_2 cap-and-trade priority:

> *So if somebody wants to build a coal-powered plant, they can. It's just that it will bankrupt them because they're going to be charged a huge sum for all that greenhouse gas that's being emitted…That will also generate billions of dollars that we can invest in solar, wind, biodiesel and other alternative energy approaches.*

See also Appendix 4-3: *EPA Mandates that New Coal Plants Prevent Nonexistent Climate Problem with Unavailable Solution*, Larry Bell, Forbes Opinions, October 1, 2013

How Bush Agencies got it Right

Bush administration agencies submitted reasons to reject a draft version of proposed EPA greenhouse gas emissions regulations under its Clean Air Act which have since been validated by reality. A July 11, 2008 policy memorandum released by the White House elaborated critical assessments expressed by leading officials from organizations that included not only the US Departments of Agriculture, Commerce, Transportation, and Energy, but even the EPA Administrator and the White House Council on Environmental Policy chair.

President Bush expressed concern in April of that year with

taking "laws written more than 30 years ago to primarily address local and regional environmental effects and applying them to global climate change. The Clean Air Act is one of these laws."

Bush warned that "If stretched beyond its original intent, it would override legislation just enacted by Congress, requiring the government to regulate far more than merely power plant emissions or cars. This would turn the Federal Government, in effect, into the Nation's local planning and zoning board, regulating countless smaller users and producers of energy—from schools and stores to hospitals and apartment buildings—with crippling effects on our economy." He said that "Decisions with such a far-reaching impact should not be left to unelected regulators and judges, but should be debated openly and made by the elected representatives of the people they affect."

Then-EPA Administrator Steve Johnson made it clear that his staff draft did not represent his organization's policy. He agreed with the president that:

> *One point is clear: the potential regulation of greenhouse gases under any portion of the Clean Air Act could result in an unprecedented expansion of EPA authority that would have a profound effect on virtually every sector of the economy and touch every household in the land.*

Johnson also believed that the Clean Air Act was "an outdated law originally enacted to control regional pollutants that cause direct health effects, is ill-suited for the task of regulating global greenhouse gases. Such a course of action would inevitably result in a very complicated, time-consuming and, likely, convoluted set of regulations."

The Department of Commerce, Energy, Transportation and Agriculture secretaries argued that applying Clean Air Act regulations to US businesses in order to address global climate change would simply export economic activity and emissions to less-regulated countries and likely not actually generate any net reduction in worldwide GHG emissions.

Commerce Secretary Carlos Gutierrez observed that reducing CO_2 emissions up to 60 percent from 2000 levels by 2050 would greatly complicate preconstruction permitting requirements for modification or new construction to large office buildings, hotels, apartment building and large retail facilities demanding "a complete assessment of the costs and benefits of such an approach."

Energy Secretary Samuel Bodman said that the EPA draft sought to "address global climate change through an enormously elaborate, complex, burdensome and expensive regulatory regime that would not be assured of significantly mitigating global atmospheric GHG concentrations and global climate change" and that such an "extraordinarily burdensome and costly regulatory program under the Clean Air Act is not the right way to go."

Transportation Secretary Mary Peters added: "It is an illusion to believe that a national consensus on climate policy can be forged via a Clean Air Act rulemaking...If implemented, the actions that the draft contemplates would significantly increase energy and transportation costs for the American people and US industry with no assurance that the regulations would materially affect global greenhouse gas atmospheric concentrations or emissions."

Bush Agriculture Secretary Ed Schafer warned "If agricultural producers were covered under such complex regulatory schemes, most (except perhaps the largest operations) would be ill-equipped to bear the costly burdens of compliance, and many would likely cease farming altogether."

Edward Lazear, Chairman, Council of Economic Advisers, and John H. Marburger, Director, Office of Science and Technology Policy wrote that actions presented in the draft "will put the United States at a competitive disadvantage, will induce economic distortions and may actually be counterproductive in reducing GHGs."

James Connaughton, Chairman, Council on Environmental Quality, summed up circumstances we are currently witnessing as the Obama administration uses the EPA to replace the will of Congress: "The staff draft employs a kitchen sink approach to the innumerable ways in which EPA would use the Clean Air Act to automatically or discretionarily regulate an unprecedented range of activities giving rise to greenhouse gas emissions."

Connaughton admonished us to recognize that such actions would effectively override the deliberate, bi-partisan decisions of elected federal and state legislatures on certain policies, and that a "case-by-case application of old regulations to an entirely new set of circumstances and parties foreshadows unrelenting confusion, conflicts over compliance, and decades-long litigation windfall for attorneys, consultants, and activists, as communities and the courts strive to sort it all out."

What's the Fracking Fuss All About?

Remember all that talk about the importance of energy security and independence from foreign oil? What if it is right there trapped in oil and natural gas-bearing shale, enough to possibly supply our US needs for centuries? Wouldn't it be a pretty good idea to use it? Well lots of people obviously don't think so. They're very determined to make sure is remains right where it is.

And where is that? It's located in 167 shale oil field deposits about a mile or more beneath the surface that are distributed among several states. According to the US Geological Survey, one known as the Green River Formation (also dubbed the "Persia of the West") may alone hold more than 1.5 trillion barrels of oil— six times the proven reserves of Saudi Arabia.

Leasing rules established in 2008 by the George W. Bush administration would have opened up about 2 million acres of that federal land in Colorado, Utah and Wyoming to leasing rules. That was before Earthjustice filed two federal lawsuits in January 2009 on behalf of 13 environmental groups to block those rules, claiming that the oil shale production techniques hadn't been perfected or reviewed, and that the royalties to be paid to government were too low.

Interior Secretary Ken Salazar then announced that the Obama administration is taking a "fresh look" at the Bush oil shale leasing rules, presumably in response to environmental lobby objections. He is well known for demonstrating kinship with their purposes, as demonstrated by actions following the Deepwater Horizon disaster in the Gulf of Mexico when all deepwater rigs were shut down while the administration took a different fresh look at safety rules and procedures.

A de facto moratorium still continues in the form of a punishingly slow permitting process. As a result, the US drilling industry in the Gulf of Mexico has collapsed, output has dropped, rigs have relocated to foreign shores, and at least one drilling company has filed for bankruptcy. This has occurred despite the fact that US District Judge Martin Feldman held that the administration had been in contempt of court for trying to reinstate it after he had already issued an injunction on grounds that such actions were too broad in scope and unjustified based upon available evidence.

Of course, our government is really only doing its job to protect us, and once again is calling upon top experts to guide policy decisions. In 2011 Energy Secretary Steven Chu announced the appointment of a seven-member panel to come up with new fracking safety standards that address concerns raised by environmental opponents. They included such domestic energy enthusiasts as former Al Gore aide and secretary of the Pennsylvania Department of Environmental Protection Kathleen McGinty, and Fred Krupp, president of the Environmental Defense Fund. This panel is remarkably similar to the one convened by Salazar that sanctioned the current offshore drilling debacle.[179]

The safety methods in question involve "fracking", a method that pumps water containing small percentages of other materials into the porous shale under pressure to hydraulically fracture it and release the trapped oil and gas. Those materials vary, and may utilize sand, nitrogen, carbon dioxide, air and other substances. Steel casings and cement prevent the fluids from entering wells and underground reservoirs. The process has been used successfully without a single documented case of site contamination over more than 60 years. More recent techniques apply horizontal drilling to extend operations into adjacent underground deposits.

These new methods are transforming oil and gas extraction productivity. The enormous Bakken Formation that extends between North Dakota (the primary area), South Dakota, eastern Montana and Canada's Saskatchewan providence is an important example. In 2005 North Dakota produced a total of about 100,000 barrels of oil per day, while today the daily output is running at 250,000 barrels from the Bakken deposit alone.

Not surprisingly, Bakken fracking has active critics who object that the process utilizes diesel fuel and solvents that can cause environmental harm. John Harju, associate director at the University of North Dakota Environmental Energy & Research Center believes this issue is exaggerated. Drillers recover 90% of these fluids, and those that aren't collected remain in oil reservoirs some 10,000 feet below the surface. He maintains that "The lowest-most occurrence of fresh water tends to appear at 2,000 feet deep or less." [180]

The biggest, most headline-winning fear is that fracking

[179] *Frack, Baby, Frack*, Investor's Business Daily, May 10, 2011
[180] *Bakken Shale Promises Big Oil Production*, Donald D. Gold, March 24, 2011, Investor's Business Daily

chemicals will contaminate drinking water. This mantra got a large popular boost following an April 19 accident at a Chesapeake Energy Corp. natural gas well in Pennsylvania about 25 miles south of the New York state border that spilled chemicals into a nearby stream, forcing evacuation of nearby residents. Workers rapidly stopped the leak, no one was injured and state authorities reported that any environmental impacts appeared to be minor.

Nevertheless, the incident heightened opposition to drilling in another massive shale deposit, a 65 million acre Marcellus Formation, the world's second largest natural gas field, extending from Ohio and West Virginia up through Pennsylvania and upstate New York. Subash Chandra, an energy analyst for Jeffries & Co. commented that "Even minor blowouts like this don't help the cause."

He was clearly correct. The Pennsylvania Department of Environmental Protection promptly sent Chesapeake a "notice of violation", the first step in enforcement proceedings. The filing asked the company to explain "why it took Chesapeake nearly 12 hours to address the uncontrolled release of fluids off the well pad" and why it took 12 hours to get a well-control company to the site. The company responded that it had a well-control specialist on the scene in 30 minutes, not 12 hours.[181][182]

Challenges to Marcellus drilling didn't begin with the Chesapeake event. Three years ago New York's powerful green lobby represented by its Department of Environmental Conservation announced plans to rewrite all fracking regulations and stop all permits, even outlawing vertical fracking. Although Governor David Peterson ultimately vetoed this, he issued an executive order that backs an agency ban on horizontal fracking,

---

[181] *The Madness of New York*, Wall Street Journal, December 16, 2010
[182] *Chesapeake Spill Heightens Pressures*, Ben Casselman, April 28, 2011, Wall Street Journal

(the most effective method) until new regulations are released.[183]

EPA convened public hearings in Binghamton, New York as part of their investigation into human health and environmental effects of fracking in March of 2010. And while the Energy Policy Act of 2005 prevents The EPA from explicitly regulating fracking wells under the Safe Drinking Water Act. This occurred after a 2004 EPA study concluded that fracking fluids posed no significant risk to drinking water. In fact both the EPA and the Ground Water Protection Council, a nonprofit made up of state regulatory agencies, have published studies determining that no documented evidence of fracking-sourced groundwater pollution has been found.

But never underestimate its willingness or ability to exercise regulatory authority under other existing laws. On August 31, the EPA quietly released interim results of its ongoing review of possible drinking water contamination at several sites near Pavillion, Wyoming. Kevin Book, an energy with ClearView Energy Partners, an energy market research firm reports, "Although EPA's latest data did not conclusively link contamination to fracking, EPA's guidance that residents should avoid drinking the water may give Congressional fracking opponents a valuable sound bite to use when calling for mandatory disclosure rules [identifying chemicals used]".[184] [185]

Halliburton, a name anti-fossil lobbies love to hate, recently announced development of a new hydraulic fracturing fluid they call "CleanStim" composed entirely of ingredients used in the food industry. Will this stem criticism? The EPA will have to study this matter, and probably the FDA also. After all, shouldn't we expect

[183] Ibid

[184] *EPA Frac Study to Focus on Water Impact*, Maureen N. Moses, July 7, 2010, AAPG Explorer

[185] *Fracking for Natural Gas: EPA Hearings Bring Protests*, Mark Clayton, September 13, 2010, Christian Science Monitor

they may have to budget for a legion of new inspectors to guard against salmonella and other food poisoning risks? As Richard Mgrdechian, author of *How the Left Was Won,* observed, "All they need to do is to hear that Halliburton is involved and they will immediately go into attack mode." [186]

We can fully expect that their continuing efforts, supported by countless petitions filed by radical environmental organizations, will lead to back-door regulation of hydraulic fracturing through the Toxic Substances Control Act, Resource Conservation and Recovery Act, and Clean Air Act, ultimately leading to restrictions upon natural gas and coal-bed methane as well.

EPA is now but one of fourteen different federal agencies that are working to find ways to regulate hydraulic fracturing in order to limit and eventually stop the practice altogether. Others include the Department of Energy (DOE), the Bureau of Land Management (BLM), the Center for Disease Control (CDC), the Department of Agriculture (USDA), and even the Securities and Exchange Commission (SEC).

Yet if fracking does indeed pose any threat, that danger may fall most directly upon Obama administration's promotion of energy independence through "renewables". Access to abundant, inexpensive fuels that contrast so starkly with uncompetitive wind and solar prospects presents good reasons for many to be alarmed. Included are "green energy" subsidy seekers and anti-fossil climate crusaders.

But no sympathy for legions of government regulatory bureaucrats is warranted. Fracking won't endanger them in the least—not so long as current policies prevail.

---

[186] *Halliburton Develops Eco-Friendly Fracking Fluid,* Brady Nelson, Environment and Climate News, March 2011

Chapter Fifteen: Green Fascism

SINCLAIR LEWIS ONCE opined that if fascism ever comes to America it will be wrapped in the flag and be carrying a cross. With regard to bureaucratic regulatory overreach it has already arrived garbed in a green flag and religious vestment.

Again, as Michael Crichton articulated the essence of this green creed in a 2003 speech:

> *There's an initial Eden, a paradise, a state of grace and unity with Nature; there's a fall from grace into a state of pollution as a result from eating from the tree of knowledge; and as a result of our actions, there is a judgment day coming for all of us. We are energy sinners, doomed to die, unless we seek salvation, which is now called sustainability. Sustainability is salvation in the church of the environment, just as organic food is its communion, that pesticide-free wafer that the right people with the right beliefs imbibe.*[187]

---

[187] Michael Crichton, *Remarks to the Commonwealth Club*, San Francisco, September 15, 2003, see:
http://www.hawaiifreepress.com/ArticlesMain/tabid/56/ID/2818/Crichton-Environmentalism-is-a-religion.aspx

As for any separation of church and the regulatory state...forget about it.

EPA-Environmental Lobby Sue and Settle Scams

"Sue and Settle" practices, sometimes referred to as "friendly lawsuits", are cozy deals through which far-left radical environmental groups file lawsuits against federal agencies wherein court-ordered "consent decrees" are issued based upon a prearranged settlement agreement they collaboratively craft together in advance behind closed doors. Then, rather than allowing the entire process to play out, the agency being sued settles the lawsuit by agreeing to move forward with the requested action they and the litigants both want.

In other words, the agency throws the case, somewhat like Br'er Rabbit agreeing to be thrown into a favorite brier-patch by Br'er Fox. A big difference in this case however, is that both were partners in the scam from the beginning. It's the unwary American public that actually does get caught in the thorns.

While the environmental group is given a seat at the table, outsiders who are most impacted are excluded, with no opportunity to object to the settlements. Accordingly, both the litigants and the defendant agency, operating in coffee bars and friendly courtroom shadows, avoid the harsh outside glare of oversight. No public notice about the settlement is released until the agreement is filed in court...after the damage has been done.

On top of all that, we taxpayers, including those impacted regulatory victims, are put on the hook for legal fees of both colluding parties. According to a 2011 GAO report, this amounted to millions of dollars awarded to environmental organizations for EPA litigations between 1995 and 2010. Three "Big Green" groups received 41% of this payback: Earthjustice, $4,655,425 (30%); the Sierra Club, $966,687; and the Natural

Resources Defense Council, $252,004. Most of this was paid to environmental attorneys in connection to lawsuits filed under the Clean Air Act, followed next by the Clean Water Act.

In addition, the Department of Justice forked over at least $43 million of our money defending EPA in court between 1998 and 2010. This didn't include money spent by EPA for their legal costs in connection with those rip-offs, since EPA doesn't keep track of their attorney's time on a case-by-case basis.[188]

While researching various broken government regulatory processes, the US Chamber of Commerce discovered many new rulemakings and unreasonable permitting delays which appeared to have resulted from such consent decrees in which the EPA agreed to bind itself to issue new regulations on a specific timetable; i.e., "We can tell Congress the court made us do it."

The Chamber concluded that such Sue and Settle rulemaking is responsible for many of EPA's "most controversial, economically significant regulations that have plagued the business community for the past few years". Included are regulations on power plants, refineries, mining operations, cement plants, chemical manufacturers, and a host of other industries. One of the most successful Sue and Settle strategies they cited "...has been on an issue few in Washington or around the nation are paying attention to: regional haze requirements under the Clean Air Act."[189]

The Chamber's study author, William Yeatman, Assistant Director for Energy and Environment at the Competitive Enterprise Institute (CEI), emphasizes that EPA's abuse of its Regional Haze authority forces states to relinquish their authority

[188] *GAO Report Exposes Millions in Environmental Litigation Fees for the First Time*, U.S. Senate Committee on Environment and Public Works: Minority Page, August 31, 2011
[189] *EPA's New Regulatory Front: Regional Haze and Takeover of State Programs*, U.S. Chamber of Commerce

and accept EPA's far more expensive plans, thereby increasing consumer utility charges. He concluded that: "...no state is immune from having its rightful Regional Haze authority trampled by EPA at profound costs for virtually nonexistent benefits."

Having conducted eight state case studies, Yeatman found:

> In Arizona, EPA's Regional Haze regulation threatens to increase the cost of water, forcing the state to spend an additional $90.2 million per year to implement the federal regulation.
>
> In Montana, EPA's proposed Regional Haze controls are nearly 250% more expensive than what that agency's standing rules presume to be "cost-effective" for compliance.
>
> In 2011, the EPA disregarded New Mexico's Regional Haze plan, instead imposing a federal plan that requires nearly $840 million more in capital costs...potentially raising average annual household utility bills by $120.
>
> Although North Dakota is one of only 12 states that achieves all of EPA's air quality standards for public health, it would not be able to achieve EPA's Regional Haze goals for visibility even by shutting down all industry. The EPA plan would also cost the state an additional $13 million per year,
>
> Refusing to approve Oklahoma's Regional Haze plan, the EPA's plan would cost the state $282 million annually.
>
> In Wyoming, the EPA proposed a federal implementation plan that would cost almost $96 million per year more than the state's plan.
>
> Minnesota is subject to back-to-back Regional Haze regulations, whereby EPA is claiming authority to regulate regional haze twice in succession at its Sherburne County generating plant.

*EPA's proposed plan would cost Nebraska nearly $24
million per year to achieve invisible "benefits".* [190]

Louisiana Senator David Vitter (R-LA), the ranking member of
the Senate Environment and Public Works Committee, plans to
investigate this Sue and Settle practice, "using all available tools to
bring to light this often abused path to regulatory influence". For
starters, he asked his Louisiana Attorney General Buddy Caldwell
to join with AGs of 13 other states who filed a Freedom of
Information Act (FOIA) request with EPA on August 10, 2012,
asking for any and all correspondence between EPA and a list of
80 environmental, labor union and public interest organizations
that have been party to litigation since the start of the Obama
administration. [191]

Unfortunately, getting federal agencies to comply with
formal FOIA requests isn't proving to be easy. And to make
matters worse, there is evidence that top bureaucrats have even
used private email accounts and aliases to cover culpability tracks.

For example, Senator Vitter, along with Representative
Darrell Issa (R-Calif.) who then headed the House Oversight and
Government Reform Committee, sent a letter to James Martin,
EPA's Region 8 administrator, warning him:

> *The use of personal, non-official email accounts raises
> concerns that you could be attempting to insulate this
> and other email correspondence from a Freedom of
> Information Act request. Moreover, your actions may also
> constitute violation of the Federal Records Act...evading*

---

[190] *U.S. Chamber Report Reveals that EPA's Takeover of States' Regional Haze
Programs is all Cost, No Benefit*, U.S. Chamber of Commerce
[191] Press Release/Letter, *Vitter Warns Louisiana of EPA's Secret 'Sue and
Settle' Deals, Could Impact State*, January 22, 2013

congressional oversight of federal agencies.[192]

One of those private emails at issue involved an exchange between Martin and Vickie Patton, the General Counsel for the Environmental Defense Fund, regarding an arranged meeting with him. Patton wrote, "Hi Jim, Next Monday or Tuesday December 12/13 at 9am depending on which is best for you." Responding, Martin replied, "January 13 at 9:00 am works for me if it works for you. (Lost your original note-is that the date and time you proposed?" Vitter and Issa have confirmed that the meeting did, in fact, take place at the EPA regional office.

Christopher Horner of the Competitive Enterprise Institute lodged a FOI lawsuit on behalf of the American Tradition Institute (ATI) seeking emails concerning the EPA's behind-the-scenes regulatory war on coal plants. Horner's book, *The Liberal War on Transparency* discusses this and other tricks.[193]

Horner also sued in the Federal District Court of Washington, D.C. to compel EPA to end its eight-month stonewall of two FOIA requests regarding its "uncomfortably close ties at great taxpayer expense" with the American Lung Association (ALA), and the Sierra Club. Both of these organizations lobby for stringent regulatory legislation the EPA wants, while at the same time, receiving agency funding.

ALA has received $20,405,655 from EPA over the past 10 years, and has run campaigns against politicians who challenge its policies. An ALA billboard ad in Michigan targeted House Energy and Commerce Committee Chairman Fred Upton, featuring a child with an oxygen mask over her face. The text read, "Rep. Fred Upton, protect our kids' health. Don't weaken the Clean Air

---

[192] *Sen. Vitter Hits the Ground Running*, Myron Ebell, February 5, 2013, Global Warming.com

[193] Christopher Horner, *The Liberal War on Transparency: Confessions of a Freedom of Information 'Criminal'*, October 2, 2012

Act."

According to the ATI complaint, "The Sierra Club employs a similar model and has close working relationships with Agency officials." This charge refers most particularly to an incident where Sierra hired disgraced EPA Region 6 Administrator Al Armendariz. He left EPA shortly after a videotape revealed him acknowledging his "philosophy of enforcement" as being akin to random crucifixions used to keep subjects suitably respectful and "really easy to manage for the next few years." Armendariz has pledged to continue to help Sierra (and, presumably, EPA) fight the coal industry.

Senator Vitter's office informed me that there are ongoing investigations regarding a known revolving door of environmentalists and agencies, as well as Equal Action to Justice Act recovery money flowing to these groups post-settlement. In other words, the litigating groups are getting paid to sue both at the front end and tail end as well.

In an affidavit, ATI informed the court that one of the two specialists EPA assigned to handle the CEI's FOIA requests admitted that a supervisor instructed her and a colleague to perform no work on them. Then, "Following this, EPA constructed a cul de sac of refusing to perform a search of responsive records until ATI agreed to pay estimated fees...which by law non-profits typically do not pay under FOIA...but which EPA then refused to provide." [194] [195]

Quite certainly, EPA is not the only federal agency to engage in secret Sue and Settle deals with crony environmental lobbying organizations. In a January letter to his Attorney General Buddy Caldwell, Senator Vitter wrote: "The collusion between federal

[194] *Public Interest Group Sues EPA for FOIA Delays, Claims Agency Ordered Officials to Ignore Requests*, The Washington Examiner, January 28, 2013
[195] Press Release, The Environmental Law Center of the American Tradition Institute, January 28, 2013

bureaucrats and the organizations entering consent agreements under a shroud of secrecy represents the antithesis of a transparent government, and your participation in the FOIA request [filed by 13 other states' AGs] will help Louisianans understand the process by which these settlements were reached." Caldwell subsequently agreed to the Senator's request.

Vitter highlighted an example where the Fish and Wildlife Service (USFWS), under management of the Department of Interior, entered an agreement with the Center for Biological Diversity involving new rules for endangered species...a deal that could impact private property owners across the South who lack financial resources necessary to fight a legal challenge from that massive federal agency.

In a massive 2011 settlement of a lawsuit filed by environmental groups, USFWS promised to address more than 250 candidate species it had previously found warranting protection under the Endangered Species Act (ESA) but were precluded from listing due to a backlog. As part of a behind-closed-door agreement, USFWS also pledged to review hundreds more species proposed for listing.[196]

Noting that this ESA-based Sue and Settle ruling presents a problem for private landowners throughout the country, Senator Vitter is also encouraging the inclusion of USFWS (which was not originally included) in the multi-state AG FOIA collusion investigation. Recognizing that the path forward towards obtaining disclosure will be difficult, Vitter added that he was, "Warily confident that both EPA and USFWS will shun all efforts to open the doors on these practices, the negotiations, and the communications between agency staff and outside groups regarding Sue and Settle agreements."

Considering that our top law enforcement official, Attorney General Eric Holder, has been cited for Contempt of Congress in

---

[196] Marita Noon, Executive Director, Energy Makes America Great, Inc.

withholding requested communications regarding the DOJ's own "Fast and Furious" debacle, there can be no lingering doubt that Senator Vitter's unfortunate prediction will prove right. Sadly, and alarmingly, this all falls far short of what would logically be expected from that "most transparent administration in history" we were promised.

See also Appendix 4-4: *EPA's Secret and Costly 'Sue And Settle' Collusion with Environmental Organizations*, Larry Bell, Forbes Opinions, February 17, 2013

The Billionaire's Club Secret Anti-Fossil War

A recent US Senate report reveals how an exclusive group of wealthy individuals directs and controls the far-left environmental movement through private foundations, which in turn influence major EPA policies centered upon restricting fossil use. Hidden under a guise of philanthropy, the foundations funnel large amounts of Big Green to intermediaries—either as a pass-through or fiscal sponsor—which transfer the money to other nonprofit 501(c)(3) and 501(c)(4) organizations the original foundations might also directly support.

The Senate Committee on Environment and Public Works Minority Staff Report stated:

> While it is uncertain why they operate in the shadows and what they are hiding, what is clear is that these individuals and foundations go to tremendous lengths to avoid public association with the far-left environmental movement they so generously fund.

Nor do they wish outsiders to know which groups they fund, or how much they give them. As noted in the report, this is all accomplished thanks to benefits of a generous tax code intended to promote genuine philanthropy and charitable acts. Unlike close attention that has been directed to Conservative non-profit organizations, this is accomplished "amazingly with little Internal Revenue Service scrutiny."

In doing so, the Billionaire's Club (the report term) "gains access to a close network of likeminded funders, environmental activists, and government bureaucrats who specialize in manufacturing phony 'grassroots' movements and promoting bogus propaganda disguised as science and news to spread an anti-fossil energy message to the unknowing public."

Donors and their recipients work in tandem to maximize the value of tax deductions while leveraging combined resources and political policy influence. This often includes lobbying on behalf of the EPA to advance policy positions important to the agency—which is statutorily prohibited from lobbying on its own behalf. In doing so, they serve as the face of the environmental movement, presenting themselves as non-partisan benevolent charities.

Meanwhile, the public remains unaware of the agency's shadowy backroom policy deals and money transfers. Successes are achieved through the "capture" of key EPA employees at the expense of farmers, miners, roughnecks, small businesses, and families. The report charges that the Obama administration is particularly guilty of deliberately staffing highest levels of its EPA with far-left environmental activists who have worked hand-in-glove with their former colleagues.

In addition to providing insider access to important policy decisions, the report accuses that the revolving door EPA then doles out large amounts of government grant money to their former employers and colleagues. More than $27 million in taxpayer—funded EPA grants have gone to major environmental groups with significant ties to senior EPA officials. The Natural Resources Defense Council and Environmental Defense Fund have each collected more than $1 million.

These awards are often premised upon propaganda disguised as science and news skillfully supplied by the Billionaire's Club and their charities. Included is alarmist anti-fracking "research" which the *Huffington Post*, *Mother Jones*, and *Climate Desk*—all grant recipients themselves—eagerly report on.

Tax loophole-savvy attorneys and accountants enable elite contributors to receive full tax benefits even when grant recipients aren't recognized as public charities and when the money indirectly funds political activities. Such relationships involve hundreds of non-profits with each set up according to a designated purpose as a cog in a well-designed machine.

This sophisticated apparatus depends heavily upon "facilitators"—both organizations and individuals—to successfully bring the private foundations and activists together. The Senate report highlights three organizations that play prominent facilitator roles: the Environmental Grantmakers Association (EGA), the Democracy Alliance (DA), and the Divest/Invest Movement.

EGA is a place where wealthy donors meet to coordinate the distribution of publicly undisclosed grants. DA facilitates broad far-left transaction agendas. (They claimed President Obama's executive actions on climate change as their success.) The Divest/Invest Movement, together with EGA, have pooled and channeled hundreds of millions of dollars to chosen activist organizations.

One of the "even more unsettling" dominant organizations mentioned in the Senate report is the Sea Change Foundation, a private California entity which relies on funding from a foreign company with undisclosed donors. Sea Change subsequently funnels tens of millions of dollars to other large foundations and prominent political activists.

The largest secretly foreign-funded Sea Change recipient is The Energy Foundation, a pass-through "public charity" utilized by EGA members to create the appearance of a more diversified support base. This is intended "to shield them from accountability and leverage limited resources by hiring dedicated

energy/environment staff to handle strategic giving."
President Obama once observed:

> *There aren't a lot of functioning democracies around the world that work in this way where you can basically have millionaires and billionaires bankrolling whoever they want, in some cases undisclosed. What it means is ordinary Americans are shut out of the process.*

Yes…and that's a subject he obviously knows a great deal about.

Chapter Sixteen: Carbon Taxomania

HOW IMPORTANT ARE those "benefits" we might expect from the EPA's war on carbon? Of 13 federal agencies invited to provide testimony about the Obama administration's anti-fossil climate policy before the House Energy Committee on September 1, 2013, two accepted, providing EPA Administrator Gina McCarthy and Energy Secretary Ernest Moniz as witnesses.

A notable exchange occurred about two hours and sixteen minutes into the hearing between McCarthy and Rep. Mike Pompeo (R-Kan.). My friend Marlo Lewis at the Competitive Enterprise Institute provided an unofficial transcribed version of the discussion segment that follows:

Pompeo: *Ms. McCarthy I want to ask a couple of questions of you. So one of the objectives today is to identify the greenhouse gas regulations that already existed and those in the future—how they actually impact the climate change, right? So you'd agree we want to have a successful climate policy as a result of those sets of rules and regulations that you promulgate? Fair base line statement?*

McCarthy: *In the context of a larger international effort, yes.*

Pompeo: *You bet. And on your website you have 26 indicators used for tracking climate change. They identify various impacts of climate change. So you would believe that the purpose of these rules is to impact those 26 indicators, right? So you put a good greenhouse gas regulation in* place, you'll get a good outcome on at least some or all of those 26 indicators.

McCarthy: *I actually...I think that the better way to think about it, if I might, is that it is part of an overall strategy that is positioning the US for leadership in an international discussion. Because climate change requires a global effort. So this is one piece and it's one step. But I think it's a significant one to show the commitment of the United States.*

Pompeo: *Do you think it would be reasonable to take the regulations you promulgated and link them to those 26 indicators that you have on your website? That this is how they impacted us?*

McCarthy: *It is unlikely that any specific one step is going to be seen as having a visible impact on any of those impacts—a visible change in any of those impacts. What I'm suggesting is that climate change [policy] has to be a broader array of actions that the US and other folks in the international community take that make significant effort towards reducing greenhouse gases and mitigating the impacts of climate change.*

Pompeo: *But these are your indicators, Ms. McCarthy. So...*

McCarthy: *They are indicators of climate change, they are not directly applicable to performance impacts of any one action.*

Pompeo: *How about the cumulative impact of your actions? Certainly you're acting in a way...you say these are indicators of climate change. Certainly it can't be the case that your testimony today is that your cumulative impact of the current set of regulations and those you're proposing isn't going to have any impact at all on any of those indicators?*

McCarthy: *I think the President was very clear. What we're attempting to do is put together a comprehensive climate plan, across the administration, that positions the US for leadership on this issue and that will prompt and leverage international discussions and action.*

Pompeo: *So you're putting regulations in place for the purpose of leadership but not to impact the indicators that you, the EPA, says are the indicators of climate change? I'm puzzled by that.*

McCarthy: *Congressman we work within the authority that Congress gave us to do what we can. But all I'm pointing out is that much more needs to be done and it needs to be looked at in that larger context.*

Pompeo: *In 2010 with NHTSA [National Highway Traffic Safety Administration], in your opening statement you said you've gotten rid of about 6 billion metric tons [of greenhouse gases]. One of your indicators, for example, is heat-related deaths. How many heat-related deaths have been eliminated as a result of the 2010 NHTSA rules?*

McCarthy: *You can't make those direct connections Congressman. Neither can I.*

Pompeo: *There's literally no connection between the activities you're undertaking and...*

McCarthy: *I didn't say that.*

Pompeo: *Well, you said you can't make the connections, so tell me what I'm not understanding. Can you draw a connection between the rules you're providing, the regulations you're promulgating, and your indicators? Or is it just...*

McCarthy: *I think what you're asking is can EPA in and of itself solve the problems of climate change. No we cannot. But the authority you gave us was to use the Clean Air Act to regulate pollution, carbon pollution is one of those regulated pollutants, and we're going to move forward with what we can do that's reasonable and appropriate.*

Pompeo: *I'm actually not asking that question that you suppose that I'm asking. I'm not asking whether you have the power to solve greenhouse gases. What I asked was: Is anything you're doing, doing any good? As measured by the indicators that you've provided. Is your testimony that you just have no capacity to identify whether the actions EPA has undertaken has any impact on those indicators? Literally, this is about science—cause and effect. Is there any causal relationship between the regulations you promulgated and the 26 indicators of climate change that you have on your website?*

McCarthy: *The indicators on the website are broad global indicators...*

Pompeo: *They're not broad, they're very specific.*

McCarthy: *...of impacts associated with climate change. They are not performance requirements or impacts related to any particular act.*

Pompeo: *I actually like the indicators—they're quantifiable,* *right? Heat-related death, change in ocean heat, sea-level rises, snow cover—those are very quantifiable things. But now you're telling me...*

McCarthy: *They indicate the public health associated with climate change.*

Pompeo: *Exactly, but you're telling me you can't link up your actions at EPA to any benefit associated with those quantifiable indicators that the EPA itself has proposed as indicative of climate change.*

McCarthy: *I think what we're able to do is to show—and I hope we will show this in the package that we put out for comment—is what kind of reductions are going be associated with our rules, what we believe they will have in terms of an economic and a public health benefit. But it is again part of a very large strategy.*

Pompeo: *My time has expired.*

See also Appendix 4-5: *EPA Head Admits Being Clueless About Any Obama Climate Plan Benefits*, Larry Bell, Forbes Opinions, September 22, 2013

So there you have it. Regardless of the countless billions of taxpayer and consumer dollars being spent to wage war on natural and inevitable climate change, the EPA head is unable to identify any discernible health and welfare benefits of her agency's draconian regulatory policies. Instead, the apparent goal of the EPA's current and proposed greenhouse gas regulations is to persuade the international community, particularly China, India, and other developing nations, to follow the Obama administration's US leadership over an economic precipice.

Nevertheless, both Dems and even some leading Republicans appear determined to curb carbon one way or another whether—costs be damned—doing so offers and measurable benefits or not.

Pushing Cap-and-Trade Using EPA

An EPA-spearheaded executive order is attempting to achieve what the Obama administration failed to accomplish through legislation. Although you have to look hard in the fine print of EPA's 645 page rule report to find it, that's where that "flexibility" so generously afforded to EPA enters the picture.

Page 39 allows states to use "strategies that are not explicitly mentioned in any of the four building blocks (e.g. market-based trading)." Page 48 says that states can combine their programs with other states. Multiple references to California's Greenhouse Gas Solutions Act (AB32) remove any lingering doubt regarding the game plan...namely encouraging states to create or combine together within a multi-state cap-and-trade system in order to meet required individual caps.

Whereas it's most likely to be referred to by the White House as a "budget program" rather than the more politically-charged "cap-and-trade" term, it will mean the same thing. The central strategy is to set emission limits on companies and entice them to trade allowances or credits as a way to stay under different benchmarks set by EPA. For example, power plant companies could either directly trade emissions credits, or use offsets in the power sector such as renewable energy or energy-efficiency programs to meet targets.

Another not-so-hidden strategy is to force states to adopt carbon taxes. Although Congress rejected a proposed national energy tax during Obama's first term, these costs will amount to just that.

Of course all states will also be pressured to install more "renewable energy" windmills and sunbeam collectors. If this isn't enough, groups such as the National Resources Defense Council and the Brookings Institute advocate a "direct load control" policy which prescribes and manages schedules when electricity users are allowed to run air conditioners and washing machines.

While these strategies will be very costly for Americans everywhere, some regions will be impacted disproportionately. Since prices for electricity are about 30 percent higher in New England and California which already require high uses of renewable fuels, it will redistribute income by placing a toll on others that don't. Particularly hard hit will be much of the South, the Ohio River Valley and mid-Atlantic.

And although coal-producing states such as Wyoming, North Dakota, Pennsylvania and Montana will get hurt the worst, they certainly aren't alone. According to a study sponsored by the US Chamber of Commerce, the shuttering of hundreds of coal-fired power plants nationwide will eliminate hundreds of thousands of jobs.

Some of those penalties fell upon Democrats from coal-intensive states that ran for reelection in 2014. Not coincidentally, some of them were among the 45 senators who sent a letter to EPA Administrator Gina McCarthy urging her to double the amount of time allowed for public comments regarding the rulings from two months to four.

There should be no doubt that the new rules will be litigated for years. The Supreme Court decision in Bond vs. US held that the federal government can make "a stark intrusion into traditional state authority" only with "a clear statement of the purpose." Yet the entire sweeping federal power grab is predicated wholly upon a short and obscure clause of the 1970 Clean Air Act (Section 111-d) predicated upon addressing real pollutants.

Then there's that Carbon Tax Thing

As much as I admire former Secretary of State George Shultz, and because I do, I was totally flummoxed by an April 2013 Wall Street Journal article he co-authored with economist Nobel laureate Gary Becker. Incredulously, the two senior fellows at Stanford University's conservative Hoover Institution expressed support for a "revenue neutral" tax on carbon.

They begin by saying "Americans like to compete on a level playing field to win on competitive merits"...then go on to argue that the way to level that playing field is to introduce that tax on what they described as a "major pollutant" to "encourage producers and consumers to shift towards energy sources that emit less carbon—such as toward gas-fired power plants and away from coal-fired plants—and generate greater demand for electric and flex-fuel cars and lesser demand for conventional gasoline-powered cars."

The op-ed piece proposes that this can be accomplished in two possible ways. It might be imposed upon the population-at-large at the point of consumption (gasoline stations and electricity bills), or alternatively, could be collected as a tax at the production level to "greatly reduce the number of collection points" through the IRS or Social Security Administration. In the case of using IRS, "an annual distribution could be made to every taxpayer and recipient of the Earned Tax Credit. In the case of SSA, the distribution could be made in terms proportionate to the dollars involved, to everyone either paying into the system or receiving benefits from it."

And you thought income redistribution was just a liberal Democrat thing?

Yeah, and the amount of the tax would then rise or fall on the basis of the subsequent increase or lessening of "climate effects" and "should be supplemented by a reasonable and sustained support for research and development in the energy area." [197]

Sadly, they aren't the only prominent and usually brilliant conservatives who seem to have sampled the carbon tax Kool-Aid. Reagan economist Arthur Laffer has said he would support such a tax in exchange for a payroll or income tax reduction; Bush 43 economist Greg Mankiw supports a global tax; and Douglas Holtz-Eakin, a senior advisor to John McCain in 2008, wants a tax to provide the energy industry with regulatory "certainty".

---

[197] *Why We Support a Revenue-Neutral Carbon Tax*, April 8, 2013, Wall Street Journal

The only certainty I foresee in such a buy-in, is a sacrifice of core conservative principles, an abandonment of common sense, and a deserved loss of political base support.[198]

Actually, we've been there...tried that! President Clinton attempted a similar gambit nearly 20 years ago with a proposed tax on energy use. His so-called "BTU tax", which would have been imposed upon every segment of the economy, provoked a brutal backlash that should have been no big surprise. He and congressional Dems soon caved in to demands of a vocal anti-BTU coalition which included small businesses, the agriculture sector, the building trades, the transportation industry, manufacturers and even social-service organizations representing poor and homeless clients who rely upon affordable gasoline and heating fuel.

This experience constituted a "teachable moment". As the great English writer Samuel Johnson observed, "There is nothing like a hanging to concentrate the mind." One year later (November 1994), Republicans took over control of the House for the first time in decades.[199]

And it's really, really stupid because the whole carbon tax idea is based upon the notion of taxing "bad" fossil energy to make it more expensive and "level the playing field" (without government picking winners and losers, of course), to make "good" energy (determined by politically-favored green-marketing crony capitalists and climate alarmists), more cost-competitive.

As my respected friend Marlo Lewis at the Competitive Enterprise Institute cogently points out, economic arguments that government should tax "bads", like CO_2 emissions, rather than "goods", like labor and capital, reflect sloppy thinking. "In

---

[198] *Carbon Tax Is Both Pointless and Inflationary*, Steve Milloy, November 14, 2012, Investor's Business Daily
[199] *Will the Carbon Tax Make a Comeback?*, William O'Keefe, December 21, 2012, Wall Street Journal

technical economic terms, only finished products and services are 'goods'. Labor and capital are inputs, production factors or costs." Here, energy is a key input without which most labor and capital would be idle or non-existent. And since about 83% of US energy comes from carbon-based fuels, a carbon tax also taxes what economists loosely call "goods". Lewis refers to this as "free-lunch economics...a recipe for failure or worse." [200]

The fundamental premise underlying the carbon tax rationale is that it offers economic compensation for harm that fossil fuel consumers inflict upon public health and the environment. Proponents argue that because fossil fuel is therefore underpriced, society uses too much of it. Therefore corrective ("Pigouvian") taxes should be added to these costs to achieve "efficient" energy markets.

This is exactly the curve being pitched by the International Monetary Fund's carbon tax proposal to provide punitive compensation for the "social cost of carbon" (SCC). And how does IMF determine that "appropriate" amount of SCC compensation? While acknowledging that estimates in the literature have varied considerably, ranging from $12 per ton [of emissions] (Nordhaus, 2011) to $85 per ton (Stern, 2006)...[IMF] estimates assume damages from global warming of $25 per ton of CO_2 emissions, following the United States Interagency Working Group on Social Cost of Carbon (2010), an extensively reviewed study.

Instead of recognizing the enormous influence that affordable fossil fuel...and yes, that includes coal...has upon virtually all aspects of our lives and economy, carbon tax proponents would have us institute policies which will deliberately undermine those benefits. At the same time we're supposed to overlook the facts that despite increased atmospheric CO_2 levels, global temperatures have been flat for going on two decades; there has

---

[200] *Why the GOP Will not Support Carbon Taxes (if it wants to survive)*, Marlo Lewis, November 26, 2012

been no increase in US flood magnitudes over the past 85 years; that long-term sea level rise is not accelerating; and that at current rates of ice loss, Greenland will contribute only about 3 inches to global sea levels by the end of this century.

If a case is made to punish carbon for its social costs and level the playing field foe alternative energy markets, shouldn't wind be taxed for environmental damage caused to those endangered birds and bats it slaughters? Shouldn't solar power be penalized for all the toxic contaminates, including heavy metals used in those solar panels that must eventually be disposed of...not to mention power plant impacts upon fragile desert ecosystems?

And then what about the wholesale land damage and water depletion that results from growing and processing all that corn for ethanol fuel which, by the way, if it mattered one whit, releases just as much atmospheric greenhouse gas as fossil fuels do? Shouldn't it be taxed as well?

Instead of leveling the playing field, let's level with the public. Do you seriously believe that a "revenue neutral" carbon tax has any chance of being enacted, or if so, staying that way for very long? For those who think that these added costs will be offset by reductions in other taxes, or that the new revenues will be returned to consumers, then go ahead. Help yourself to some more Kool-Aid.

For the rest of you, expect that a newly imposed carbon tax would be revenue neutral only briefly, at best, and then become but one more cookie jar to raid in futile attempts to keep up with escalating demands of government's spending addiction. And if you harbor any illusions that those tax rates won't increase over time, consider a lesson taken from European VAT experience. Since the time it was implemented, the initial VAT rate of 10% in Germany has climbed to 19%, in France it is up from 13.6% to 19.6%, and in several other countries VAT rates now exceed

20%.[201]

From a social cost perspective, carbon taxes are, by nature, regressive, meaning that they inflict largest pain burdens upon low-income households. As Marlo Lewis observes, this presents a "Catch-22" dilemma for any Republicans. If, on one hand, they offer to support a carbon tax in exchange for cuts in corporate or capital gains taxes, they will be accused of seeking to benefit the rich at the expense of the poor. On the other hand, any "carbon dividends" paid out to offset higher energy price burdens on poor households will create a new class of welfare dependents…a costly consequence for the general public. This is something that Democrats are much less inclined to worry about.

Finally, the carbon tax concept is inherently parasitic, a strategy that attacks the vitality of the economic host upon which it feeds. Such "playing field-leveling" penalties, piled on top of other ever-expanding anti-carbon regulatory disincentives, would inhibit growth of one of the few bright spots in the US economy, the development of our vast coal, oil and natural gas deposits.

Such policies, driven by unfounded climate alarmism, motivated by green crony capitalism, and embracing the vision of transforming America into a European-style socialist welfare state, should not be countenanced by any authentic Republican. For the party to do so would not only impose unacceptable social costs, but deservedly high political costs as well.

See also Appendix4-6: *Carbon Tax…Are Republicans Really That Stupid?*, Larry Bell, Forbes Opinions, April 16, 2013

So here's an alternate idea. Rather than further penalizing households and businesses with a pointless and painful carbon tax, why not do just the opposite? Let's impose an anti-carbon tax on profiteering prophets of the climate alarm industry that gives them something to really worry about.

---

[201] Ibid

Section 5

Truly Scary UN Agendas

WE HAVE BEEN led to be frightened about the wrong human-caused climate disaster. The one truly worthy of alarm is the "Climate of Corruption: Politics and Power Behind the Global Warming Hoax" headlined in the title of my previous book.

Are "corruption" and "hoax" unduly harsh terms for characterizing political agenda-driven manipulations of science? Don't just take my words for it. Selective climate data and doctored graphs in IPCC's 1996 *Summary for Policymakers* report which also featured text changes which were made after the scientists had approved it and before it was printed prompted the late Dr. Frederick Seitz, a world-famous physicist and former president of the US Academy of Sciences, the American Physical Society and the Rockefeller University to write in the Wall Street Journal:

> *I have never witnessed a more disturbing corruption of the peer review process than events that led to this IPCC report.*

Contrived climate crisis rubric has provided an ideal platform to accomplish exactly what UN doom and gloom drama producers and IPCC directors have in mind...namely to advance large transformational visions of socialism, wealth redistribution, and ultimately, global governance.

If this sounds a bit too conspiratorial, again consider the words spoken by former Soviet Union President Mikhail Gorbachev recognizing the importance of using climate alarmism to advance socialist Marxist objectives, stating in 1996, "The threat of environmental crisis will be the international disaster key to unlock the New World Order." This may well have seemed like the last hope for that agenda following the USSR's economic and political collapse in 1991.[202]

Also, once more think about words of a speech delivered by then-President Jacques Chirac of France supporting a key Western European Kyoto Protocol objective, "For the first time, humanity is instituting a genuine instrument of global governance, one that should find a place within the World Environmental Organization which France and the European Union would like to see established."[203]

IPCC Summary for Policymakers reports offer repeated prescriptions to carry out UN agendas aimed most particularly at redistributing American wealth in penance for our unfair capitalist free market prosperity. Included are well orchestrated initiatives that promote regionalized (smaller) economies to reduce transportation demand, achieve resource-sharing through co-ownership, and encourage citizens to pursue free time over personal wealth.

As IPCC official Ottmar Edenhofer admitted in November 2010, "...one has to free oneself from the illusion that

[202] Sovereign Independent
http://www.sovereignindependent.com/?p=18097
[203] *Climate Talks or Wealth Redistribution Talks?*, Nicolas Loris, November19, 2010, Heritage.org

international climate policy is environmental policy. Instead, climate change policy is about how we redistribute de facto the world's wealth..."

See also Appendix 5-1: *Yes! We Should Defund The UN's Intergovernmental Panel On Climate Change!*, Larry Bell, Forbes Opinions, February 24, 2013

Chapter Seventeen: Historical Hype, Hijinks and Hysteria

FOR HISTORICAL CONTEXT, consider that the recent crisis-premised eco-fascism movement draws scripting from an earlier disaster drama. For if there is one person to be attributed the title *Father of Manipulated Gloom and Doom Environmental Fright*, it must be Thomas Robert Malthus, a political economy professor at the British East India Company's East India College who lived from 1766-1834. His "zero-sum-gain" population and resource theories have had tremendous influence on global agendas, policies and travesties which continue unabated today.

Malthus initiated an alarmist international movement with an unsigned pamphlet titled *An Essay on the Principle of Population* that first appeared in London bookstores in 1798. The publication forecast a terrifying world future whereby the population would increase geometrically while agriculture necessary to sustain it would increase only arithmetically.

Malthus proclaimed as "incontrovertible truths" that because of the "fixity of land", growing families would overwhelm means to feed them. This circumstance would lead to "misery or vice"— some combination of disease, famine, foregone marriage, barbarianism and war that reduced population to a sustainable subsistence level. This, he argued, would be "decisive against the existence of a society, all the members of which should live in ease, happiness, and comparative leisure."

The remedies Malthus proposed to ensure lives of "ease, happiness and comparative leisure" were draconian to say the least. For example, he argued to condemn doctors who find cures in order to reduce population...even encouraged efforts to keep wages low:

> We are bound in justice and honor to disclaim the right of the poor to support...[W]e should facilitate, instead of foolishly and vainly endeavouring to impede, the operations of nature in producing mortality; and if we dread the too frequent visitation of the horrid form of famine, we should sedulously encourage the other forms of destruction, which compel nature to use. Instead of recommending cleanliness to the poor, we should encourage contrary habits.

Malthus went on to propose:

> In our towns we should make streets narrower, crowd more people into the houses, and court the return of the plague. In the country, we should build our villages near stagnant pools, and particularly encourage settlements in all marshy and unwholesome situations. But above all, we should reprobate specific remedies for ravaging diseases; and the benevolent, but much mistaken men, who have thought they were doing a service to mankind by projecting schemes for the total extirpation of particular disorders.

Even during his own time, his theories were used to justify regressive legislation against lower classes...influences that led to establishment of England's Poor Law Act of 1834 and which motivated the British government to refuse aid during the Irish Famine of 1846.

Yet also during Malthus's time, while England's population was growing, the food supply was actually growing even more rapidly. Studies were soon beginning to indicate an inverse relationship between wealth and population change, where wealthier regions had lower growth rates. Then with the birth of the European Industrial Revolution, living conditions for many improved dramatically, if not by standards approaching those we enjoy today.

Human Population as a Virus

The 1971 book, *Global Ecology*, authored by Paul Ehrlich and John Holdren (now Obama administration "Science Czar") presented an unmistakable and continuing Malthusian theme:

> When a population of organisms grows in a finite environment, sooner or later it will encounter a resource limit. This phenomenon, described by ecologists as reaching a 'carrying capacity' of the environment, applies to bacteria on a culture dish, to fruit flies in a jar of agar, and to buffalo on a prairie. It must also apply to man on this finite planet.

This was followed by a 1974 book by Paul and Anne Ehrlich, *The End of Affluence: a Blueprint for Your Future*, where they predict:

> In the early 1970s, the leading edge of the age of scarcity arrived. With it came a clearer look at the future, revealing more of the nature of the Dark Age to come.

Mankind at the Turning Point (1974), a book published by the Club of Rome, an elitist international organization, presented humankind as a disease to be eradicated. It said, "The World Has Cancer and the Cancer Is Man". Another book, *The Population Bomb Revisited*, written by Paul Ehrlich and his wife Anne (2009), explains:

> *A cancer is an uncontrolled multiplication of cells; the population explosion is an uncontrolled multiplication of people...We must shift our efforts from the treatment of symptoms to the cutting out of the cancer. The operation will demand many apparently brutal and heartless decisions. The pain may be intense.*

An earlier Ehrlich book, *The Population Bomb*, which followed Hugh Moore's pamphlet of the same title in 1968, said:

> *The battle to feed all of humanity is over. In the 1970s and 1980s hundreds of millions of people will starve to death...At this late date nothing can prevent a substantial increase in the world death rate...Nothing could be more misleading to our children than our present affluent society. They will inherit a totally different world, a world in which the standards, politics and economics of the past decade are dead.*

It goes on:

> *Our position requires that we take immediate action at home and promote effective action worldwide. We must have population control at home, hopefully through changes in our value system, but by compulsion if voluntary methods fail. Americans must also change their way of living so as to minimize their impact on the world's resources and environment.*

In his book, Ehrlich proposed that a "Federal Bureau of Population and Environment should be set up to determine the optimum population size for the United States, and devise measures to establish it." One suggestion was that the bureau might add "temporary sterilants to water supplies or staple food. Doses of the antidote would be carefully rationed by the government to produce the desired population size."

Anticipating some political opposition to such a strategy, Ehrlich suggested a backup plan that involved revising the US tax code:

> On top of the tax change, luxury taxes could be placed on layettes, cribs, diapers, diaper services, expensive toys...There would, of course, have to be considerable experimenting on the level of financial pressure necessary to achieve the population goals. To the penalties could be added some incentives.

For example, he suggested that:

> A government 'first marriage grant' could be awarded each couple in which the age of both partners was 25 or more. 'Responsibility prizes' could be given to each couple for each five years of childless marriage, or to each man who accepted irreversible sterilization [vasectomy] before having more than two children. Or special lotteries might be held—tickets going only to the childless.

With help from television, radio and newspaper publicity organized by the Sierra Club, *The Population Bomb* was a best seller.

Elite members of the internationally influential Club of Rome expressed similarly jaded views in their 1991 publication

Chapter Seventeen: Hype, Hijinks and Hysteria

The First Global Revolution:

> *In searching for a new enemy to unite us, we came up with the idea that pollution, the threat of global warming, water shortages, famine and the like will fill the bill...But in designating them as the enemy, we fall into the trap of mistaking symptoms for causes. All these dangers are caused by human intervention and it is only through changed attitudes and behavior that they can be overcome. The real enemy then, is humanity.*

In other words, "We have met the enemy, and it is us."

In June, 1978, Song Jian, a top-level manager in charge of developing control systems for the Chinese guided missile program presented the premise regarding limited "carrying capacities" presented in the Club of Rome's earlier *Limits to Growth* and *Blueprint for Survival* publications to Deng Xiaoping and other leaders of the Communist Party. The idea resonated as an argument to blame poverty on population rather than 30 years of misrule. Song then had an instrumental role in preparing a 100 year plan, with a one child per family policy to take effect immediately. Ehrlich's dream of compulsory birth control had finally been put into practice.

In a 1967 speech at a University of Texas symposium (quoted in *The Legacy of Malthus: The Social Costs of the New Scientific Racism*), Ehrlich called for the US to shift research from "short-sighted" medical programs aimed at keeping people alive, to population regulation...and in particular, let an overcrowded Indian population "slip down the drain"...requiring sterilization of men with three or more children in India as a condition for food aid.

> *We should change the pattern of federal support of biomedical research so that a majority of it goes into broad areas of population regulation, environmental*

sciences, behavioral sciences, and related areas, rather than short-sighted programs on death control. It is absurd to be preoccupied with the medical quality of life until and unless the problem of the quantity of life is solved...India, where population growth is colossal, agriculture hopelessly antiquated, and the government incompetent, will be one of those we must allow to slip down the drain.

President Eisenhower had patently rejected the US-funded or mandated international population control, stating at a December 2, 1959 press conference, "I cannot imagine anything more emphatically a subject that is not a proper political or government activity or function or responsibility...That is not our business." Then-Senator John F. Kennedy, who was running for president at the time, opposed such programs as well...a stand that drew stinging attacks from his Democrat party challengers who accused him of trying to impose his Catholic values.

Planned Parenthood leader Margaret Sanger declared that she would leave the country if Kennedy was elected in 1960. (He was...she didn't.) In 1961 her Planned Parenthood Federation merged with Moore's World Population Emergency Campaign to form the Planned Parenthood-World Population Society.

Yet after President Kennedy's assassination this anti-population control stance would change, and by the mid-1960s a new government policy doctrine was official. The US Agency for International Development (USAID) not only provided population control assistance to Third World countries, but the Johnson administration pointedly refused to provide famine relief to India following 1966 crop failures unless its government agreed to impose forced sterilization programs upon its rural peasantry.

Faced with mass starvation the requirement was implemented...and having participated in promoting this policy, Secretary of Defense Robert McNamara who served under

Kennedy and Johnson, left the administration to head the World Bank.

As Robert Zubrin notes in his book *Merchants of Despair*:

> *Collectively, these entities—the Population Council, the Draper Fund/Population Crisis Committee, [Sanger's] International Planned Parenthood Federation, USAID, the World Bank, and the UN Fund for Population Activities (largely funded by USAID)—together with a host of smaller outfits funded by them (including the Population Reference Bureau, the Association for Voluntary Sterilization, the Pathfinder Fund, and many others, would come to form the imposing and influential population control establishment.*

See also Appendix 5-2: *Books: Robert Zubrin's Merchants of Despair Reveals Racism and Genocide Cloaked in Green Camouflage*, Larry Bell, Forbes Opinions, July 31, 2013

Imagine that control establishment's disciple and Ehrlich colleague, John Holdren, would ultimately come to be selected from all of our nation's countless possibilities to become President Obama's Science Czar.

Capitalism in the "De-Development" Cross Hairs

Perhaps unremarkably, it was that very same John Holdren who wrote in the introduction to *Global Ecology* which he coauthored with Paul Ehrlich:

> *Only one rational path is open to us—simultaneous de-development of the [overdeveloped countries] and semi-development of the underdeveloped countries (UDC's), in*

206

order to approach a decent and ecologically-sustainable
standard of living for all in between. By de-development
we mean lower per-capita energy consumption, fewer
gadgets, and the abolition of planned obsolescence.

In their book *Population, Resources, and Environment*, Holder and
coauthors Paul and Anne Ehrlich repeat this theme, stating, "The
need for de-development presents our economists with a major
challenge. They must design a stable, low-consumption economy
in which there is a much more equitable distribution of wealth
than in the present one. Redistribution of wealth both within and
among nations is absolutely essential if a decent life is to be
provided for every human being."

Holdren and the Ehrlichs put their anti-growth philosophy
into a mathematical equation where a negative environmental
impact is correlated with a combination of population growth,
increasing affluence, and improving technology. In an effort to
minimize environmental damage, their solution proposes
"...organized evasive action: population control, limitation of
material consumption, redistribution of wealth, transitions to
technologies that are environmentally and socially less disruptive
than today's, and movement toward some kind of world
government." [204]

Apparently unaware that CO_2 fertilizes vegetation, Holdren
predicted in 1986 that "carbon dioxide-induced famines could kill
as many as a billion people before the year 2020." Then in 2006 he
suggested that global warming could cause world-wide sea levels
to rise by 13 feet by the end of this century, whereas in reality,
the sea levels have been rising at the rate of about seven inches per
century since the end of the Little Ice Age. And while there are

---

[204] Paul Ehrlich, Anne Ehrlich, and John Holdren, *Ecoscience: Population,*
Resources, and Environment, (San Francisco: W. H. Freeman and
Company, 1977), p. 954

some various local areas where sea levels are rising and falling, there is no recent global rate of acceleration.[205][206]

In the October 2008 issue of *Scientific American*, Holdren wrote:

> *The ongoing disruption of the Earth's climate by man-made greenhouse gases is already well beyond dangerous and is careening toward completely unmanageable.*

Carbon dioxide, he added…

> *…is the most important of civilization's emissions and the most difficult to reduce. About 80 percent comes from burning coal, oil and natural gas; most of the rest comes from deforestation in the tropics.*

There should be little wonder then that science advisor Holdren must certainly have vetted the climate statements that President Obama offered in his January 2014 State of the Union address where he referred to "carbon pollution" three times. This was presented in connection with his commitment to "set new standards on the amount of carbon pollution our power plants are allowed to dump into the air."

The EPA now mandates a 1,100-pound limit per megawatt hour on carbon emissions from new coal power plants. This is not only despite the lack of any scientifically-supportable evidence of climate benefit, but also the fact that there is no viable commercial-scale technology to achieve that ideological

---

[205] Paul Ehrlich, *The Machinery of Nature*, Simon & Schuster, New York, 1986, p. 274

[206] *Obama's Top Science Adviser to Congress: Earth Could Be Reaching Global Warming 'Tipping Point' That Would Be Followed by a Dramatic Rise in Sea Level*, December 9, 2009 , CNS News

pipedream. As previously mentioned - even if it mattered, the most modern coal-fired plants can only reduce CO_2 emissions to 1,800 pounds. What's more, the White House and EPA already knew that.

Chapter Eighteen: UN Plans for America

CLIMATE GUILT-MONGERING offers an ideal strategy to extract American wealth in penance for our unfair capitalist free market prosperity. If you have any doubt about this, perhaps some events that took place during the UN's 2009 Copenhagen, Denmark climate conference might be illuminating.

First, Bolivian President Evo Morales addressed the delegates stating:

> The real cause of climate change is the capitalist system. If we want to save the earth then we must end that economic model. Capitalism wants to address climate change with carbon markets. We denounce those markets and the countries which [promote them]. It's time to stop making money from the disgrace that they have perpetrated.

Then, after developing countries demanded that rich ones provide many billions of dollars to them for damage to the climate, US and European representatives expressed willingness to provide their "fair share", pledging $10 billion per year from 2010 to 2012. This offer was rejected as an insufficient insult, representatives of several undeveloped countries walked out of the meetings and angry riots broke out in the streets.

Secretary of State Hillary Clinton then came to the rescue, offering to up the ante with a $100 billion annual contribution from the United States and our more prosperous friends to the "poorest and most vulnerable [nations] among us" by 2020. Where it would actually come from no one knew, including Hillary and her then-boss. (Any guesses?)

Yet judging from the tumultuous standing ovation following a speech from then-Venezuelan president Hugo Chavez, one might have imagined that he was going to provide all that money. But this was not so. Instead he had aroused general agreement in the audience regarding where to lay the blame for the world's social, economic and climate problems, "Our revolution seeks to help all people...Socialism, is the other ghost that is probably wandering around this room, that's the way to save the planet; capitalism is the road to hell...Let's fight against capitalism and make it obey us."

Viewed from an even larger perspective, the global warming crisis-premised anti-capitalism rubric has provided an ideal platform to accomplish exactly what Chavez has in mind...to enable the UN to advance large transformational visions of socialism, wealth redistribution, and ultimately, global governance.

Activist groups attending a Venezuela government-hosted, UN-backed environmental conference made it clear that going green isn't about preserving green forests, or even about capping plant-nourishing carbon emissions for a greener planet in order to halt climate change. Nope, the "Margarita Declaration" handed down by the 130 environmental ministers who attended the four-day July 2014 meeting called carbon-capping markets a "false solution" to the problem of climate change and branded the UN's forest conservation scheme "dangerous and unethical". Instead, the real solution is to cap-and-trade capitalism for socialism.

As agreed in a run-up to the UN's main round of climate talks in Lima, the blame was clear, "The structural causes of climate change are linked to the current hegemonic [capitalist] system...To combat climate change it is necessary to change the system."

Transforming the Global Economy (Starting with Ours)

One year after former US Secretary of State Hillary Clinton pledged that $100 billion annually by 2020 to help Third World countries address climate problems attributed to America and capitalism at the UN-sponsored 2009 World Climate-Change Summit in Copenhagen, a 2010 Cancun, Mexico meeting that followed focused primarily upon a formulating a plan of action to implement that commitment. This commenced literally with a vengeance, producing the design for a new $100 billion per year "Green Climate Fund" (GCF). Its purpose is nothing less than to fundamentally transform the global economy...beginning with ours.

And the key players? They include George Soros, a World Bank representative, and Lawrence Summers, former director of the White House National Economic Council and assistant to President Obama for economic policy. Be advised to take what is envisioned very seriously.

The first GCF meeting of the 40-member design team, the "Transitional Committee" (TC), took place in Mexico City in April to prepare "operational specifications" for the GCF in time for approval at a later UNFCCC meeting in Durban, South Africa. This fell right in line with proposals prepared by a 20-member "High-Level Advisory Group on Climate Change Financing" assembled by UN Secretary-General Ban ki-Moon for the purpose of calling upon nations "to fundamentally transform the global

economy-based on low-carbon, clean energy resources."
Resulting recommendations are to institute:

- Carbon tax
- Tax on international aviation and shipping
- Financial transaction tax
- Wire tax for producing electricity
- Redirection of 100% of all fossil fuel subsidies for international climate action.[207 208]

Fox News reported possession of position papers indicating topics covered in meetings involving Ban ki-Moon and 60 of his top lieutenants during a two-day 2010 Australian Alpine Retreat discussing ways to put their sprawling organization in charge of world agendas. Those topics included:

> *How to restore "climate change" as a top global priority after the Copenhagen fiasco*
>
> *How to continue to try to make global redistribution of wealth the real basis of that climate agenda, and widen the discussion further to encompass the idea of "global public goods"*
>
> *How to keep growing UN peacekeeping efforts into missions involved in police, courts, legal systems and other aspects of strife-torn countries*
>
> *How to capitalize on the global tide of immigrants from poor nations to rich ones, to encompass a new "international migration governance network"*

---

[207] *Green Climate Fund Fundamentally Transforms the Global Economy*, Cathie Adams, April 29/ May 6, 2011, Eagle Forum
[208] *Climate Change is Not About the Environment, But About Redistributing Wealth*, Conference of the Parties 16, Cathie Adams, Eagle Forum

How to make "clever use of technologies to deepen direct ties with what the UN calls "civil society", meaning novel ways to bypass its member nation states and deal directly with constituencies that support UN agendas

An underlying position paper theme those top UN bosses grappled with was how to cope with the pesky issue of national sovereignty that interferes with their global governance goal. As one paper put it, "the UN should be able to take the lead in setting the global agenda, engage effectively with other multinational and regional organizations as well as civil society and non-state stakeholders, and transform itself into a tool to help implement the globally agreed objectives." It went on to state, "it will be necessary to deeply reflect on the substance of sovereignty, and accept that changes in our perceptions are a good indication of the direction we are going." [209]

And just how much of this sovereignty sabotage has to do with sinking capitalism, stifling consumption and achieving global wealth redistribution? A bunch.

As an opening session explained: "The real challenge comes from the exponential growth of the global consumerist society driven by ever higher aspirations of the upper and middle layers in rich countries as well as the expanding demand of emerging middle-class in developing countries. Our true ambition should be therefore creating incentives for the profound transformation of attitudes and consumption styles." A paper prepared by Secretary General Ban's own climate change team called for "nothing less than a fundamental transformation of the global economy." [210]

Such ambitious goals will of course require lots money to carry out...our money, along with new taxing authority to steal

---

[209] *After a Year of Setbacks, UN Looks to Take Charge of World's Agenda*, George Russell, September 8, 2010, Fox News
[210] Ibid

Larry Bell

it. And manufactured carbon-based climate change alarmism
provides an excuse to get it.

In 2004 the United Nations University-World Institute for
Development Economics Research (UNU-WIDER) published a
study addressing possible scenarios for implementing a global tax.
It asked "How can we find an extra US$550 billion for
development funding? Our focus is on flows of resources from
high-income to developing countries."

The conclusion?

> *Any foreseeable global tax will be introduced, not by a
> unitary world government, but as the result of concerted
> action of nation states...The taxation of environmental
> externalities is an obvious potential source of revenue.*

It went on to say:

> *We are presupposing that the tax is indeed levied on
> individuals and firms in the form of a carbon levy.*

Another UNU-WIDER publication states: "Support for an
international 'carbon tax' has been growing since the 1992 UN
Earth Summit focused international attention on the damage to
the environment caused by excessive use of fossil fuels
worldwide...Over 20% of the tax yields would originate in the
US." [211]

As Ottmar Edenhofer, a German economist and co-chair of
the IPCC Working Group III on Mitigation of Climate Change
admitted in an *Investor's Business Daily* interview, "The climate
summit in Cancun at the end of the month [December, 2010] is
not a climate conference, but one of the largest economic

---

[211] *The History of the Global Warming Scare*, July 28, 2009, Global
Warming Science

conferences since the Second World War." [212]

Neither the Cancun meeting or the Copenhagen conference that preceded it paid any attention to revelation of science frauds associated the "ClimateGate" and "GlacierGate" scandals (the latter involving a fabricated IPCC report that Himalayan glaciers were melting due to man-made global warming). Instead, all seemed much more interested in creating the Green Climate Fund to replace many smaller agreements between nations that have developed after the 1997 Kyoto Protocol terms were established.

Agenda 21 Has Probably Already Come to Your Neighborhood

Let's flash back again even earlier for some historical perspective. Roughly 15 years after many "scientific experts" had warned about the arrival of another Ice Age, an estimated 35,000 government officials, diplomats, Non-Government Organization (NGO) activists and journalists from 178 countries attending a UN-sponsored 1992 Conference on Environment and Development (UNCED, or "Earth Summit") in Rio de Janeiro, Brazil, began to negotiate international agreements to counter a reverse climate threat. An observed shift to warming was seized upon as the basis for stabilizing "dangerous" anthropogenic greenhouse gases (principally CO_2) at 1900 levels.

By that time they had already established a Framework Convention on Climate Change (FCCC), and determined that "human activities have been substantially increasing the atmospheric concentrations of greenhouse gasses, that these increases enhance the natural greenhouse effect, and that this will result on average in the additional warming of the Earth's surface and atmosphere and may adversely affect natural ecosystems and

---

[212] Investors.com

http://news.investors.com/ArticlePrint.aspx?id=554439

mankind."

The Rio Earth Summit codified the UN's central theme for the famous (or infamous) Kyoto Protocol, which the US has refused to ratify. Along with the UN's Framework Convention on Climate Change (UNFCCC) which established the IPCC, it also produced three other much lesser-known initiatives—the Rio Declaration on Environment and Development, the Statement of Forest Principles, The United Nations Convention on Biological Diversity, and Sustainable Development Agenda 21. The latter may ultimately prove over the long term to be most broadly influential of all.

Earth Summit chairman Maurice Strong left no doubt about where to place blame for global problems, stating in the conference report:

> It is clear that current lifestyles and consumption patterns of the affluent middle class...involving high meat intake, consumption of large amounts of frozen and convenience foods, ownership of motor vehicles, golf courses, small electric appliances, home and work place air-conditioning, and suburban housing are not sustainable...A shift is necessary toward lifestyles less geared to environmentally damaging consumption patterns.

Saving the planet makes for a great cover story to conceal power and wealth redistribution that have no such laudable purpose as was witnessed at the 2009 Copenhagen Summit followed by Cancun in 2010. This organized movement really originated in the 1970s and early 1980s when Third World countries, by force of numbers, and European socialist green parties, through powers of aggressiveness, seized control of the UN and began calling for a New International Economic Order (NIEO). Continuing goals are to transfer unfair wealth from the industrialized West to their majority; to establish global socialism; and to obtain postcolonial reparations for our past misdeeds.

Now, enter "Sustainable Development Agenda 21" (or just Agenda 21). Realizing difficulties in getting some national governments—the US in particular—to enact massive regulatory control over virtually all aspects of energy production and consumption, UNCED leaders launched a campaign to advance the same agenda through state and local initiatives. Since Agenda 21 was a "soft-law" policy recommendation, (not a treaty) it required no US congressional authorization. President Clinton established a President's council in 1993 on Sustainable Development by Executive Order for the specific purpose of implementing it in the US.

The Agenda 21 plan was hatched in 1990 through an NGO called the "International Council for Local Environmental Initiatives" (ICLEI). Its name was changed in 2003 to "ICLEI—Local Governments for Sustainability" to emphasize "local" and diminish concerns about "international" influence and associations with UN political and financial ties. The ICLEI's are in most of our US counties today.

That camouflage is pretty transparent. On its web page "ICLEI: Connecting Leaders" the organization explains that their networking strategy connects cities and local governments to the United Nations and other international bodies. In doing so it represents local governments at the UN Commission on Sustainable Development, the UNFCCC and UN Conventions on Biodiversity and Combating Desertification. It also cooperates with the UN Environment Programme and UN-HABITAT II Programme. Habitat II is the implementation plan for Agenda 21.

Agenda 21 envisions a global scheme for healthcare, education, nutrition, agriculture, labor, production, and consumption. A summary version titled *AGENDA 21: The Earth Summit Strategy to Save Our Planet* (Earthpress, 1993), calls for:

...a profound reorientation of all human society, unlike anything the world has ever experienced—a major shift in the priorities of both governments and individuals and an unprecedented redeployment of human and financial resources.

The report emphasizes that:

This shift will demand a concern for the environmental consequences of every human action be integrated into individual and collective decision-making at every level.

And don't think for a moment that this will come about without lots of pain. As Al Gore stated in his 1993 book *Earth in the Balance*, sustainable development will bring about "a wrenching transformation" of American society. Come to think of it, he does appear a bit stressed lately—with all those big homes to sustain.

ICLEI's web page states that its Local Agenda 21 [LA21] Model Communities Programme is "designed to aid local governments in implementing Chapter 28 of Agenda 21, the global action plan for sustainable development." As Gary Lawrence, a planner for the city of Seattle and an advisor to the Clinton-Gore administration's Council on Sustainable Development and to US AID commented at a 1998 UN Environmental Development Forum in London titled *The Future of Local Agenda 21 in the New Millennium*, "In some cases, LA21 is seen as an attack on the power of the nation-state."

He went on to say, "Participating in a UN-advocated planning process will very likely bring out many...who would work to defeat any elected official...undertaking Local Agenda 21...So we will call our process something else, such as comprehensive planning, growth management or smart growth."

And so they have. "Comprehensive planning", "growth management" and "smart growth" (which is Agenda 21 with a new name). All mean pretty much the same thing...centralized control of virtually every aspect of urban life: energy and water use, housing stock and allocation, population levels, public health and dietary regimens, resources and recycling, "social justice" and education.

Richard Rothschild, a candidate for Carroll County commissioner in Maryland, described reasons to stave off smart growth overreach initiatives there:

> *Smart growth is not science; it is political dogma combined with an insidious dose of social engineering...It urbanizes rural towns with high-density development, and gerrymanders population centers through the use of housing initiatives that enable people with weak patterns of personal financial responsibility to acquire homes in higher-income areas...shifting the voting patterns of rural communities from Right to Left.*

Every county in America now has sustainable development directives guided by federal agencies, NGOs, and/or the ICLEI. ICLEI-USA claims that its official members consistently rank among the greenest cities and "...have led the effort in recent years to envision, accelerate and achieve strong climate protection goals, creating cleaner, healthier, more economically viable communities." More than 130 of those members are California counties and cities. They include all major metro areas—Los Angeles, Sacramento, Santa Cruz, San Diego, and San Francisco—as well as smaller communities from Alameda to Yountville. So much for influences on economic viability!

And as for goals of "sustainability", what, according to the ICLEI definition, does this term actually mean? Former Earth Summit chair, now president of the UN's University for Peace council Maurice Strong provided an answer in the forward he wrote for the *Local Agenda 21 Planning Guide* published by ICLEI, the International Development Research Center (IDRC) and UN Environment Programme (UNEP) in 1996:

> *The realities of life on our planet dictate that continued economic development as we know it cannot be sustained...Sustainable development, therefore, is a program of action for local and global economic reform—a program that has yet to be defined.*

In other words, it means whatever the United Nations ultimately decides is best for all of us.[213]

---

[213] *More UN Insanity We are Paying For: Isn't it Crazy to Continue?*, Larry Bell, Forbes Opinions, December 6, 2011

Chapter Nineteen: Time to Defund UN-Sponsored Insanity

IF THE OBAMA administration's annual pledge of $100 billion to assist other countries in solving fictitious climate problems premised upon the UN's politically-corrupted science-driven agenda, there's an even more fundamental question.

Why do we continue to support the UN's Framework Convention on Climate Change that sponsors IPCC at all, knowing that its activities and conclusions are rife with manipulation and outright corruption?

Why do we blithely continue to enable and empower the UN's Agenda 21 plan, operating through its UNFCCC, UN Commission on Sustainable Development, UN Environment Program and other organizations in the form of an International Council for Local Environmental Initiatives (ICLEI) to advance a global scheme for healthcare, education, nutrition, agriculture, labor, production and consumption within counties and communities throughout America?

Why would we tolerate the UN's planned abridgement of our Second Amendment rights as former Secretary of State Hillary Clinton announced the Obama administration's intent to work hand-in-glove with them to pass a new "Small Arms Treaty" disguised as an "International Arms Control Treaty" which is intended to force America to:

> *Enact tougher gun licensing requirements that create more bureaucratic red tape*
>
> *Confiscate and destroy all "unauthorized" civilian firearms*
>
> *Ban the trade, sale and ownership of all semi-automatic weapons*
>
> *Create an international gun registry, setting the stage for future confiscation.*[214]

Enough Already!

US Representative Blaine Luetkemeyer (R-MO) had a great idea when he introduced legislation to discontinue any more taxpayer green from being used to advance the UN's economy-ravaging agendas. Strongly objecting to the UNFCCC's use of IPCC's suggestions and faulty data to implement a job-killing agenda here in America, he argues:

> *The American people should not have to foot the bill for an international organization that is fraught with waste, engaged in dubious science, and is promoting an agenda that will destroy jobs and drive up the cost of energy in the United States. Unfortunately, the president appears to be ready to fund these groups, revive harmful policies like cap and trade, and further empower out of control federal regulators at a time when we should be doing everything possible to cut wasteful spending, reduce regulatory red tape, and promote economic growth.*

Some members of Congress favor broader accountability measures

---

[214] *UN Agreement Should Have All Gun Owners Up in Arms*, Larry Bell, Forbes Opinions, June 7, 2011

which would require that all future US funding for all UN programs comply with responsible policies. For example in 2011 House Foreign Affairs Committee Chairwoman Ileana Ros-Lehtinen introduced a United Nations Transparency, Accountability and Reform Act of 2011 (H.R. 2829) which would make US funding voluntary rather than based upon an assessment subject to their cession of anti-American and anti-Israel policies. The plan would have eliminated our contributions to the Palestinian Authority through the UN Relief and Works Agency (UNRWA) which has been caught bankrolling terrorism.[215]

That same strategy might be applied to cut off funding for future UN climate conferences aimed at ginning up IPCC junk science-spawned hysteria to bankroll their war on American capitalism. Such a measure, of course, would be strongly opposed by those who believe we must continue to subsidize the UN if we want to be popular and respected by the international community. But there's just one really big question.

Is any of that working yet?

Some Chilling Developments for UN Alarmists

On a brighter note, following nearly two decades of flat global mean temperatures maybe the UN is finally witnessing some evidence of a global climate change after all. Their own man-made crisis was un-ushered in during the organization's December 2012 Doha, Qatar meeting in the form of a typhoon named Christopher Monckton, the third viscount of Benchley, advisor to former Prime Minister Margaret Thatcher, and climate realist. Temporarily purloining a vacant microphone assigned to a delegate from Burma, he gave the entire audience some very terrifying news...announcing that "in the 16 years we have been

---

[215] *An Accountable UN At Long Last?*, September 1, 2011, Investor's Business Daily

coming to these conferences, there has been no global warming at all." [216]

If that wasn't scary enough, amid boos and heckles, Monckton blasted the congregation, uttering the utter blasphemy:

> *If we are to take action [the sort they always propose], the cost of that would be many times greater than the cost of taking adaptive measures later. So our recommendation, therefore, is that we should initiate very quickly a review of the science to make sure we are all on the right track.*

Yes, you read that right. He had the audacity to question the "science" behind the UN's money grab demands.

As Lord Monckton pointed out once again in September 2014, satellite remote sensing data shows, "there has been no global warming—none at all—for at least 215 months." He observed:

> *This is the longest continuous period without any warming in the global instrument data since satellites first watched in 1979. It has endured for half the satellite temperature record. Yet the Great Pause coincides with a continuing, rapid increase in atmospheric CO_2 concentration.*

Monckton continued:

> *The Great Pause is a growing embarrassment to those who had told us with 'substantial confidence' that the science was settled and the debate over. Nature has other*

---

[216] *The UN's Global Warming War on Capitalism: An Important History Lesson*, Larry Bell, Forbes Opinions, January 22, 2013

ideas.[217]

Let's also take a Great Pause from that hype and hysteria and recognize just as everything humans have done in the name of industrial progress hasn't been kind to the environment, it shouldn't be assumed that everything is bad either. Yet the modern-day environmental activist movement tends not to see it that way.

If there is perceived climate change, then it must be for the worst and our fault. Since human activities including industry create pollution, then most inputs and products of those endeavors, including CO_2 emissions that enable plants to grow, must be pollutants. If some glaciers appear to be melting more rapidly than they were a few decades ago, then we must be responsible, and the condition dire.

Yes, have no doubt about it—climate change is real. It will continue to happen, for better and worse without our influence or permission. The Industrial Revolution didn't start it, nor did economic progress and prosperity afforded by free market capitalism which will enable civilization to adapt.

Let's finally recognize that there are big differences between responsible stewardship ideals that most of us subscribe to, and ideologically moralistic, anti-development obstructionists who use fear and guilt to exert costly and unchecked influence over ever-expanding aspects of our liberties and lives.

Who are these prophets of doom, and how do they profit? They are compromised climate scientists and sponsors who recognize that when fear goes away, research funding will follow.

They are activist environmental lobbies who use alarmist pseudoscience claims to raise donations which influence special interest legislation.

---

[217] *British Researcher: No Global Warming in 18 Years*, Drew MacKenzie, September 9, 2014, NewsMax

They are overreaching regulatory agencies and bureaucrats who are in league with eco-zealots to conduct back-door sue-and-settle scams which apply junk science arguments.

They are "green" energy marketers who rely upon those unsupportable arguments, environmental activists, and their shared regulatory cronies to gain preferential tax and rare-payer subsidies.

They are politicians who engage in "save the planet" fear-mongering and cozy deals with anti-fossil lobbies to gain votes and fill election coffers.

They are global green movement UN delegates working through their control of FCCC/IPCC to attack capitalism while simultaneously working to redistribute its fruits of prosperity among failed nation regimes.

And it is influential agents of socialism who seize upon contrived climate and environmental crises to undermine constitutionally guaranteed American rights including private property ownership and ultimately achieve world government dominion.

Is this collaboration of interconnected, interdependent agendas winning?

So far, yes. And this constitutes a real man-caused crisis we will be witless to ignore.

Notes

Section 1: The Climate Alarm Industry

(1) Kevin Trenberth, *Predictions of Climate*, Climate Feedback: The Climate Change Blog, June 2007 http://blogs.nature.com/climatefeedback/2007/08/predictions _of_climate.html

(2) Robert Berner and Zawert Kothavala, *GeoCarb III: A revised Model of Atmospheric CO_2 over Phanerzoic Time*, American Journal of Science, vol. 301 (February 2001)

(3) Ian Wishart, AirCon, *The Seriously Inconvenient Truth about Global Warming*, HATM Publishing, 2009, 35

(4) *MIT Researcher Finds Evidence of Ancient Climate Swings*, Science Daily, April 20, 1998

(5) Ernst-George Beck, *180 Years of Atmospheric CO_2 Gas Analysis bt Chemical Methods*, Energy and Environment, vol. 18, no, 2 2007

(6) Roy W. Spencer, *Climate Confusion*, Encounter Books 2008, 87

(7) S. Fred Singer and Dennis T. Avery, *Unstoppable Global Warming*, Rowman &Littlefield, 2007, 138

(8) US Environmental Protection Agency, Global Warming-Climate, October 14, 2004

(9) Syun-Ichi Akasofu, *Two Natural Components of Recent Climate Change*, March 30, 2009,

www.webcommentary.com/docs/2natural.pdf

(10) Singer and Avery, Ibid

(11) Graeme Stephens, *Cloud Feedbacks in the Climate System: A Critical Review*, Journal of Climate. Vol. 18, January 15, 2005, 237-73

(12) *Peer Reviewed Study Rocks Climate Debate! Nature Not Man Responsible for Recent Global Warming*, Climate Depot, July 22, 2009

(13) Roy Spencer, John Christy, et al., *Cirrus Disappearance: Warming Might Thin Heat-Trapping Clouds*, University of Alabama-Huntsville, August 9, 2007
http://uah.edu/news/newsread.php?newsID=875

(14) *Climate Forecasting Models: Not Pretty—Not Smart*, Larry Bell, Forbes Opinions, August 9, 2011
http://www.forbes.com/sites/larrybell/2011/08/09/climate-forecasting-models-arent-pretty-and-they-arent-smart/

(15) IPCC Climate Change 2001, Chapter 8. *Model Evaluation*, 475
http://www.grida.no/climate/ipcc_tar/wg1/pdf/TAR-08.pdf

(16) Climate Change Reconsidered; The Report of the Nongovernmental Panel on Climate Change, 2009, p 9.

(17) IPCC, 2007-I, p 591

(16) Climate Change Reconsidered; The Report of the Nongovernmental Panel on Climate Change, 2009, p 9.

(17) IPCC, 2007-I, p 591

(18) Trenberth quote, K. Trenberth, *Predictions of Climate*, Climate Feedback: The Climate Change Blog, June 2007,
http://blogs.nature.com/climatefeedback/2007/06/predictions_of_climate.html

(19) Syun-Ichi Akasofu, *Two Natural Components of the Recent Climate Change*, March 30, 2009
www.webcommentary.com/docs/2natural.pdf

(20) Graeme Stephens, *Cloud Feedbacks in the Climate System: a Critical Review*, Journal of Climate, vol. 18 (January 15, 2005)

237-73

(21) *Pioneer Meteorologist Unearthed Mysteries of Clouds, Storms*, Wall Street Journal, March 2, 2010.

(22) Roy W. Spencer, *Climate Confusion* (Encounter Books, 2008), 87

(23) *Peer-Reviewed Study Rocks Climate Debate! Nature Not Man Responsible for Recent Global Warming*, Climate Depot, (July 22, 2009).

(24) Climate Change Reconsidered; The Report of the Nongovernmental Panel on Climate Change, 2009, p 10-1.

(25) Green, K.C. and Armstrong, J.S. 2007. *Global Warming Forecasts by Scientists Versus Scientific Forecasts*, Energy Environ, 18: 997-1021

(26) Energy Tribune, *Is It Really The Warmest Ever?*, Joseph D'Aleo, January 29, 2011
http://www.energytribune.com/articles.cfm/6440/Is-It-Really-The-Warmest-Ever

(27) Sources (Email Quotations): *Climategate 2: The Scandal Continues*, Myron Ebell, CEI Press Release
http://noconsensus.wordpress.com/2011/11/22/climategate-2-0/
http://globalwarming.org/wp-content/uploads/2011/11/FOIA2011.zip

(28) *If You See Something, Say Something*, Michael Mann, January 17, 2014, New York Times,
http://www.nytimes.com/2014/01/19/opinion/sunday/if-you-see-something-say-something.html?_r=0

(29) *Mike's Nature Trick*, Watts Up With That, November 20, 2009,
http://wattsupwiththat.com/2009/11/20/mikes-nature-trick/

(30) *Michael Mann And The ClimateGate Whitewash: Part One*, Larry Bell, June 28, 2011, Forbes,
http://www.forbes.com/sites/larrybell/2011/06/28/michael-mann-and-the-climategate-whitewash-part-one/

(30) *Michael Mann And The ClimateGate Whitewash: Part One*, Larry Bell, June 28, 2011, Forbes, http://www.forbes.com/sites/larrybell/2011/06/28/michael-mann-and-the-climategate-whitewash-part-one/

(32) *The Climategate Inquiries*, Global Warming Policy Foundation, Andrew "Bishop Hill" Montford

(33) *ClimateGate Whitewash*, S. Fred Singer, April 14, 2010 American Thinker

(34) United Nations Framework Convention on Climate Change: 2002 http://unfccc.int/resource/docs/convkp/conveng.pdf

(35) *Greens Real Target: US Economy*, Investor's Business Daily, December 8, 2009 http://investors.com/NewsAndAnalysis/Article-aspx?id=514610

(36) National Center Dossier http://www.nationalcenter.org/dos7130.htm

(37) Zbigniew Jaworowski, *CO_2: The Greatest Scientific Scandal of Our Time*, EIR Science, March 16, 2007, p.3, 16 http://www.warwickhughes.com/icecore/zjmar07.pdf

(38) Quoted by Terence Corcoran, *Global Warming: The Real Agenda*, Financial Post, 26 December 1998, the Calgary Herald, December, 14, 1998.)

(39) Sovereign Independent http://www.sovereignindependent.com/?p=18097

(40) *Climate Talks or Wealth Redistribution Talks?*, Nicolas Loris, November 19, 2010, Heritage.org,

(41) Roy W. Spencer, *Climate Confusion*, (Encounter Books, 2008, p 147)

(42) Jonathan Schell, *Our Fragile Earth*, Discover (October 1989); page 44.

(43) *Hot Talk: Cold Science*, Page 8, S. Fred Singer, The Independent Institute

(44) IPCC 1996A:5, *Are Human Activities Contributing to*

Climate Change?, http://www.gcrio.org/ipcc/qa/03.html.

(45) IPCC 2001D6, *A Report of Working Group I of the Intergovernmental Panel on Climate Change*, 10., http://www.ipcc.ch/pdf/assessment report/ar4/wg1/ar4/wg1-spm.pdf.

(46) Trenberth quote, Kevin Trenberth, *Predictions of Climate*, Climate Feedback: The Climate Change Blog, June 2007, http://blogs.nature.com/climatefeedback/2007/06/predictions_of_climate.html

(47) Lawrence Solomon, *The Deniers, Part III—The Hurricane Expert Who Stood Up To UN Junk Science*, National Post, February 2, 2007.

(48) Steven Mosher and Thomas Fuller, *Climategate: The Crutape Letters*, Vol 1, CreateSpace, 2010.

(49) *When Scientists Become Politicians*, Investors Business Journal, December 1, 2009.

(50) Sources (Email Quotations): *Climategate 2: The Scandal Continues*, Myron Ebell, CEI Press Release
http://noconsensus.wordpress.com/2011/11/22/climategate-2-0/
http://globalwarming.org/wp-content/uploads/2011/11/FOIA2011.zip

(51) *Environmentalist Icon Says He Overstated Climate Change*, Investor's Business Journal, April 25, 2012,
http://news.investors.com/article/609042/201204241902/gaia-theorist-admits-errors-on-climate-change.htm

(52) *The Compelling Case Against Ed Davey*, Melanie Phillips, February 7, 2012, The Daily Mail.com,
http://www.dailymail.co.uk/debate/article-2098014/The-compelling-case-Ed-Davey.html?ITO=socialnet-twitter-maildebate

(53) *A Top German Environmentalist Cools on Global Warming*, Larry Bell, February 14, 2012, Forbes,
http://www.forbes.com/sites/larrybell/2012/02/14/a-top-

german-environmentalist-cools-on-global-warming/

(52) *The Compelling Case against Ed Davey*, Melanie Phillips, February 7, 2012, The Daily Mail.com,
http://www.dailymail.co.uk/debate/article-2098014/The-compelling-case-Ed-Davey.html?ITO=socialnet-twitter-maildebate

(53) *A Top German Environmentalist Cools On Global Warming*, Larry Bell, February 14, 2012, Forbes,
http://www.forbes.com/sites/larrybell/2012/02/14/a-top-german-environmentalist-cools-on-global-warming/

(54) Physics Review Papers
http://www.atmosp.physics.utoronto.ca/people/vyushin/Papers/Govindan_Vyushin_PRL_2002.pdf

(55) *German Fear of Warming Plummets...Yet-To-Be-Published Skeptic Book Climbs To Amazon.de No.4!*, P. Gosselin, January 30, 2012, NoTricksZone
http://notrickszone.com/2012/01/30/german-fear-of-warming-plummets-yet-to-be-published-skeptic-book-climbs-to-amazon-de-no-4/

(56) *69% Say It' Likely Scientists Have Falsified Global Warming Research*, August 3, 2011, Rasmussen Reports,
http://www.rasmussenreports.com/public_content/politics/current_events/environment_energy/69_say_it_s_likely_scientists_have_falsified_global_warming_research

(57) *Climate of Fear*, Nature Editorial, March 11, 2010,
http://www.nature.com/nature/journal/v464/n7286/full/464141a.html

(58) *The Machinery of Nature*, Simon & Schuster, New York, 1986, p. 274.
http://www.amazon.com/Human-Ecology-Paul-Holdren-Ehrlich/dp/B001K571OM

(59) *Senators Spar During Hearing Over Alleged 1970s Cooling Consensus*, Daily Caller, March 3, 2011,
http://dailycaller.com/2011/03/03/senators-spar-during-

hearing-over-alleged-1970s-global-cooling-
consensus/#ixzz1tYipJ0ux

(60) *Environment Greenpeace Founder Questions Man-Made Global Warming*, Jonathan M. Seidl, January 20, 2011, The Blaze, http://www.theblaze.com/stories/greenpeace-founder-questions-man-made-global-warming/

(61) *Climate of Fear*, Nature Editorial, March 11, 2010, http://www.nature.com/nature/journal/v464/n7286/full/464141a.html

(62) *The Alarming Cost of Climate Change Hysteria*, Larry Bell, Forbes Opinions, August 23, 2011
http://www.forbes.com/sites/larrybell/2011/08/23/the-alarming-cost-of-climate-change-hysteria/

(63) Roy W. Spencer, *Climate Confusion*, (Encounter Books, 2008), 98-99

(64) *Ivar Giaever: Scientists Debunk Global Warming*, Geoff Metcalf, Newsmax.com, December 15, 2008
www://Newsmax.com/Metcalf/global_warming_hype.2008/12/15/161919.html

Section 2: Disastrous Speculations

(1) *When the Earth Refuses to Warm*, Wesley Pruden, January 31, 2012, The Washington Times,
http://www.washingtontimes.com/news/2012/jan/31/pruden-when-the-earth-refuses-to-warm/

(2) *Little Change in Global Drought over the Past 60 Years*, Justin Sheffield, Eric F. Wood and Michael L. Roderick, November, http://www.nature.com/nature/journal/v491/n7424/nature11575/metrics

(3) New York Times: *Climate Changes Called Ominous*, June 19, 1975—Harold M. Schmeck,—p. 31.
http://ruby.fgcu.edu/courses/twimberley/EnviroPhilo/Ominous.pdf

(4) New York Times: *Scientists Ask Why World Climate is*

Changing, Major Cooling May Be Ahead, May 21, 1975——By Walter Sullivan
http://ruby.fgcu.edu/courses/twimberley/EnviroPhilo/Why.pdf

(5) Time Magazine: *Another Ice Age*, June 24, 1974
http://www.time.com/time/printout/0,8816,944914,00.html

(6) *The Cooling World*, Newsweek, Gwynne,
http://www.denisdutton.com/cooling_world.htm

(7) 1977 book, *The Weather Conspiracy: The Coming of the New Ice Age——CIA Feared Global Cooling*
http://icecap.us/images/uploads/Excerpts_from_the_August_1977_book.pdf

(8) July 9, 1971 Washington Post article presented in the Washington Times, September 19, 2007
http://www.washingtontimes.com/news/2007/sep/19/inside-the-beltway-69748548/

(9) *Stephen Schneider Greenhouse Superstar*, John L. Daly
http://www.john-daly.com/schneidr.htm

(10) *The Faithful Heretic: A Wisconsin Icon Pursues Tough Questions*, The Wisconsin Energy Conservative
http://www.weenmagazine.com/2007issues/may/may07.html #1

(11) *Scientists Considered Pouring Soot Over the Arctic in the 1970s to Help Melt the Ice——In Order to Prevent Another Ice Age*, December 17, 2009, Washingtonsblog

(12) *Arctic Ocean Warming, Icebergs Growing Scarce Washington Post Reports*, Kirk Myers, March 2, 2010, Examiner.com,
http://www.examiner.com/seminole-county-environmental-news-in-orlando/arctic-ocean-warming-icebergs-growing-scarce-washington-post-reports

(13) Braithwaite, R.J. 2002, *Glacier Mass Balance: The First 50 Years of International Monitoring*, Progress in Physical Geography 26: 76-95

(14) Michael Asher, *So Much for Flooded Cities; Greenland Ice*

Loss Not Increasing, Daily Tech, July 4, 2008.
http://www.dailytech.com/article.aspx?newsid=12277

(15) Ian Howat, Ian R. Joughin, and Ted A. Scambos, *Rapid Changes in Ice Discharge From Greenland Glaciers*, Science, February 8, 2007.
http://www.sciencemag.org/cgi/content/abstract/1138478vl.

(16) *A Reconciled Estimate of Ice-Sheet Balance*, Science,
http://www.sciencemag.org/content/338/6111/1183

(17) *Grim Picture of Polar Ice-sheet Loss*, Olive Hefferman, November 29, 2012, Nature/News,
http://www.nature.com/news/grim-picture-of-polar-ice-sheet-loss-1.11921

(18) "New science upsets calculations on sea level rise, climate change ice sheet melt massively overestimated, satellites show", Lewis Page, November 28, 2012, The Register,
http://www.theregister.co.uk/2012/11/28/sea_levels_new_science_climate_change/

(19) *Princeton Geoscientists Report Greenland Ice Sheet Melting Rate is Increasing*, Bob Yirka, November 20, 2012
http://phys.org/news/2012-11-princeton-geoscientists-greenland-ice-sheet.html#jCp

(20) *Recent Contributions of Glaciers and Ice Caps to Sea Level Rise*,
http://sealevel.colorado.edu/content/recent-contributions-glaciers-and-ice-caps-sea-level-rise

(21) *New Sat Data Shows Himalayan Glaciers Hardly Melting at All: Results 'really were a surprise', say Climate Profs*, Lewis Page, February 9, 2012, The Register,
http://www.theregister.co.uk/2012/02/09/grace_data_himalayas_not_melting/

(22) Church, J.A., White, N.J., Coleman, R., Lambeck, K. and Mitrovica, J.X., 2004, *Estimates of the Regional Distribution of Sea Level Rise Over the 1950-2000 Period*, Journal of Climate 17: 2609-2625.

(23) Cazenave, A. and Nerem, R.S., 2004, *Present-Day Sea Level Change: Observations and Causes*, Reviews of Geophysics, 42: 10.1029/2003RG000139.

(24) Douglas, B.C., 1991, *Global Sea Level Rise*, Journal of Geophysical Research, 96: 6981-6992.

(25) Douglas, B,C., 1992, *Global Sea Level Acceleration*, Journal of Geophysical Research,97: 12,6999-12,706.

(26) Carton, J.A., Giese, B.S., and Grodsky, S.A., 2005, *Sea Level Rise and the Warming of the Oceans in Simple Ocean Data Assimilation (SODA) Ocean Reanalysis*, Journal of Geophysical Research 110: 10.1029/2004JC002817.

(27) *The State's Sea Level Retreat*, Orrin H. Pilkey, February 23, 2012, News and Observer, http://www.newsobserver.com/2012/02/23/v-print/1876250/the-states-sea-level-retreat.html

(28) Commentary: *Sea Level Reports Need Clarifying*, Bob Emory, February 26, 2012, Sun Journal, http://www.newbernsj.com/articles/crc-104665-reports-level.html

(29) *Ocean Acidification Rate May Be Unprecedented, Study Says*, Ira Block, March 1, 2012, National Geographic, http://earth.columbia.edu/articles/view/2951

(30) *Climate Change Reconsidered*, Science and Environmental Policy Project 2009 Interim Report, Heartland Institute

(31) Caldeira, K. et al., 2003, *Anthropogenic carbon and Ocean pH*, Nature

(32) Buddemeie, R.W., et al., *Coral Reefs & Global Climate Change: Potential Contributions of Climate Change to Stresses on Coral Reef Ecosystems*, The Pew Center on Global Climate Change.

(33) De'ath, G. et al., 2009, *Declining coral calcification on the Great Barrier Reef*, Science, 323

(34) Tans, P. 2009, *An Accounting of the Observed Increase in Oceanic and Atmospheric CO_2, and the Outlook for the Future*, Oceanography 22

(35) Pelejero, C et al., 2010, *Paleo-perspectives on Ocean Acidification*, Trends in Ecology and Evolution 25.

(36) Middleboe, A.L, and Hansen, P.J., 2007, *High pH in Shallow-water Macroalgal Habitats*, Marine Ecology Progress Series 338

(37) Suwa, R., et al, 2010, *Effects of Acidified Seawater on Early Stages of Scleractinian Corals (genus Acropora)*, Fisheries Science.

(38) Krief, S., et al, 2010, *Physiological and Isotopic Responses of Sceleratinian Corals to Ocean Acidification*, Geochimica et Cosmochimica Acta 74.

(39) Ries, J.B., et al, *Marine Calcifiers Exhibit Mixed Responses to CO$_2$-induced Ocean Acidification*, Geology 37.

(40) Hendriks, I.E., et al, 2010, *Vulnerability of Marine Biodiversity to Ocean Acidification: A Meta-analysis*, Estuarine, Coastal and Shelf Science.

(41) Gary Sharp, *Coral Bleaching: What (or Who) Dunnit?*, Technology Commerce Society Daily, April 26, 2006 www.tcsdaily.com/article.aspx?id=42606B

(42) The Washington Times, E. Calvin Beisner, June 4, 2010 http://www.washingtontimes.com/news/2010/jun/4/move-over-global-warming/print/

(43) *Ten Years to Solve Nature Crisis, UN Meeting Hears*, Richard Black, October 18, 2010 BBC News: Science and Environment www://www.bbc.co.uk/news/science-environment-11563513

(44) Nathan Burchfiel, *Polar Bear Scene Could Maul Energy Production*, May 7, 2008. Available at: www.BusinessAndMedia.org/printer/2008/20080507104256.aspx.

(45) Scott Whitlock, *ABC's Sam Champion Hypes Global Warming for Eight Minutes*, NewsBusters.org March 28, 2008. Available at: http://newsbusters.org/blogs/scott-whitlock/2008/03/28/abc-sam-champion-hypes-global-warming.

(46) *Interior Secretary Kemthorne Announces Proposal to List Polar Bears Under Endangered Species Act*, US Department of Interior, Press Release, December 26, 2006. Available at:
http://www.doi.gov/archive/news/06_News_Releases/061227.html.

(47) *Global Warming Link to Drowned Polar Bears Melts Under Searing Fed Probe*, Audrey Hudson, August 8, 2011, Human Events, Posted by James Pat Guerrero, August 13, 2011, Wordpress

(48) *Polar Bear Population Growth Confounds Libs*, April 6, 2012,
http://nation.foxnews.com/global-warming/2012/04/06/polar-bear-population-growth-confounds-libs#ixzz1sEfxXBWU

(49) *Emperor Penguins are Teeming in Antarctica*, Robert Lee Hotz, April 14, 2012, Wall Street Journal
http://online.wsj.com/article/SB1000142405270230362400457734185268577323 4.html

(50) *More Interior Scientists Are Taking Heat*, Felicity Barranger, September 21, 2011, New York Times
http://green.blogs.nytimes.com/2011/09/21/more-interior-scientists-are-taking-heat/

(51) Tom DeWeese, *Stupid Human Tricks: The Sad Case of the Spotted Owl*, July 2, 2007
http://www.intellectualconservative.com/2007/07/02/stupid-human-tricks-the-sad-case-of-the-spotted-owl

(52) *Blasting Some Owls to Save Others?*, Nancy Grace, CBS News, April 27, 2007 CBS.com Stories
http://www.cbsnews.com/2100-205_162-2736996.html

(53) 2000 Documentary, *Amazon Rainforest: Clear-Cutting the Myths*,
http://www.climatedepot.com/a/1846/Climate-Depot-Rainforest-Factsheet-ClearCutting-the-Myths—About-the-Amazon-and-Tropical-

Rainforests&sa=U&ei=31dxTa_EHI13UgAfcOPxP&ved=OCBE
QFJAB&usg=AFQjCNGiUYKTW_uGQ
(54) Climate Depot/ Yahoo News
news.yahoo.com/s/afp/20110304/ts_afp/environmentbiodivers
ityextinction_20110304055833
(55) Paul Reiter: *Global Warming Won't Spread Malaria*, EIR
Science & Environment (April 6, 2007: 52-57.

Section 3: Colorblind "Green Energy" Madness

(1) Industrial Wind Action Group, *Transmission Issues Associated with Renewable Energy in Texas*, March 28, 2005
http://www.windaction.org.documents/15707
(2) ERCOTT: Tudor, Pickering, Holt & Co., Energy Investment and Banking, *Texas Wind Generation Report*, 2009
http://www.tudorpickens.com/pdfs/THP.Texas.Wind.Generat
ion.Report.2009.pdf
(3) National Renewable Energy Laboratory, *Improving Wind Turbine Gearbox Reliability*, W. Musial, S. Butterfield and B. McNiff, May 7-10, 2007
http://www.nrel.gov/wind/pdfs/41548.pdf
(4) Investor's Business Daily, *The Big Wind-Power Cover-Up*, March 12, 2010
http://www.investors.com?NewsAndAnalysis?ArticlePrint.aspx?
aspx?id+527214
(5) Los Angeles Times, *First US Offshore Wind Energy Project Faces Lawsuit*, Tribune Washington Bureau, Kim Geiger, June 26, 2010
http://articles.latimes.com/print/2010/jun/26/nation/la-na-
wind-turbines-20100626
(6) *Solar Power Showing Greater Mainstream Potential*, October 23, 2011, Chicago Sun-Times
http://www.suntimes.com/news/nation/8380521-418/solar-
power-showing-greater-mainstream-potential.html
(7) *Sustainability: Some Free Marker Reflections*, Marlo Lewis,

February 11, 2011, MasterResource
http://www.masterresource.org/2011/02/sustainability-post/

(8) *Renewables Are Unsustainable*, Paul Driessen, CFACT

(9) *Solar Power is Beginning to Go Mainstream*, Jonathan Fahey, Associated Press/ USA Today,
http://www.usatoday.com/money/industries/energy/story/20 11-10-30/solar-power-energy/50979764/1

(10) *Wind Power Subsidies? No Thanks*, Patrick Lenevein, April 1, 2013, Wall Street Journal,
http://online.wsj.com/article/SB100014241278873235010045 78386501479255158.html

(11) *Power Struggle: Green Energy Versus a Grid That's Not Ready*, Evan Halper, December 2, 2013, *LA Times*,
http://www.latimes.com/nation/la-na-grid-renewables-20131203,0,1019786.story#axzz2mjO8I33K

(12) Global Wind Energy Council
http://www.gwec.net/fileadmin/images/Publications/7Top10c umulativecapacityDec2010.jpg

(13) *The Wind Experience*, Institute for Energy Research
http://www.instituteforenergyresearch.org/2011/05/09/the-wind-experience

(14) *Britain Faces Winter of Blackouts as Firms Are Asked to Ration Electricity*, Donna Rachel Edmunds, September 3, 2014, 1,
http://www.breitbart.com/Breitbart-London/2014/09/03/Britain-Faces-Winter-of-Blackouts-As-Firms-Are-Asked-to-Ration-Electricity

(15) *Misguided Energy Policies Have Put Europe on a Path to Economic Decline*, Larry Bell, City AM newspaper (London), October 11, 2013
http://www.cityam.com/article/1381452108/misguided-energy-policies-have-put-europe-path-economic decline?utm_source=homepage_puff&utm_medium=homepage_ puff&utm_term=pufftest&utm_campaign=homepage_puff

(16) *Lessons from Lemmings: The E.U.'s Green Power Folly*, Larry

Bell, Forbes Opinions, October 3, 2011

(17) *Undoing America's Ethanol Mistake*, Senator Kay Bailey Hutchinson, April 28, 2008,

http://hutchinson.senate.gov/opedEthanol.html

(18) Boston Herald/AP, *The Secret Environmental Costs of US Ethanol Policy*,

http://bostonherald.com/news_opinion/national/2013/11/the
_secret_environmental_cost_of_us_ethanol_policy

(19) *Ethanol Isn't Worth Costlier Corn Flakes and Tortillas*, Michael Economides, May 17, 2011, Forbes.com

(20) *How Biofuels Could Starve the Poor*, Benjamin Senauer, Foreign Affairs May/June 2007

(21) *Ethanol Comes with Environmental Impact, Despite Green Image*, Tom Davies, May 5, 2007, USA Today http://www.usatoday.com/money/industries/environment/20 07-05-o5-ethanolenvironment_N.html

(22) *Biofuels May Emit More Greenhouse Gas*, Matt Ball, February 9, 2008 Vector 1 Mediz.com,

http://www.vector1media.com/spatialsustain/biofuels-may-emit-more-greenhouse-gas.html

(23) *Water Use by Ethanol Plants*, Dennis Keeney and Mark Muller, October 2006, Institute for Agriculture and Trade Policy www.iatp.org/iatp/publications.cfm?accountID=258&refID=89 449

(24) *Ethanol vs. Water: Can Both Win?*, Sea Stachura, September 18, 2006 Minnesota Public Radio

http://minnesota.piblicradio.org/display/web/2006/09/07/eth anolnow

(25) *Ethanol vs. Water: Can Both Win?*, Sea Stachura, September 18, 2006 Minnesota Public Radio

http://minnesota.piblicradio.org/display/web/2006/09/07/eth anolnow

(26) *EPA's Ethanol Mandate for 2014 Behind Schedule*, Christopher Doering, Des Moines Register, Monday, June 30,

2014
http://www.desmoinesregister.com/story/money/agriculture/2014/06/27/epa-ethanol-mandate-weeks-away-gas-supply-congress-renewable-fuel-standard/11447021/

(27) *EPA Reduces 2014 Ethanol Mandate*, James Taylor, the Heartlander, December 24, 2014,
http://news.heartland.org/newspaper-article/2013/12/23/epa-reduces-2014-ethanol-mandate

(28) Producers *Panic as Ethanol Mandate Loses Support*, James Stafford, Free Republic, January 2, 2014
http://www.freerepublic.com/focus/news/3107823/posts?page=1

(29) *Ethanol Added $14.5 Billion to Consumer Motor Fuel Costs in 2011*, Study Finds, Marlo Lewis, July 19, 2012, GlobalWarming.com
http://www.globalwarming.org/2012/07/19/ethanol-added-14-5-billion-to-consumer-motor-fuel-costs-in-2011-study-finds/

(30) *The RPS, Fuel and Fuel Prices, and the Need for Statutory Flexibility*, Thomas E. Elam, July 16, 2012, FarmEcon LLC,
http://www.globalwarming.org/wp-content/uploads/2012/07/RFS-issues-FARMECON-LLC-7-16-12.pdf

(31) Coordinating Research Council Report,
http://www.crcao.com/reports/recentstudies2012/CM-136-09-1B%20Engine%20Durability/CRC%20CM-136-09-1B%20Final%20Report.pdf

(32) *Auto Industry Adopts New Warning for Gas Cap Labels*,
http://www.nmma.net/assets/cabinets/Cabinet421/E15%20cap%20label.JPG

(33) *EPA Approves E15 Fuel Label Despite Engine Risk*, James R. Healy, June 29, 2011, USA Today,
http://content.usatoday.com/communities/driveon/post/2011/06/e15-e10-ethanol-alcohol-pollution-engine-damage-labels-gas-station-pump-epa/1

(34) *Can Boutique Fuel Save Small Engines from the Wear and Tear of E10*, Roy Berendsohn, August 12, 2011, Popular Mechanics,
http://www.popularmechanics.com/home/reviews/outdoor-tools/can-boutique-fuel-save-small-engines-from-the-wear-and-tear-of-e10

(35) *Ethanol Mandate Waiver: Decks Stacked Against Petitioners*, Marlo Lewis, September 10, 2012, Global Warming,
http://www.globalwarming.org/2012/09/10/ethanol-mandate-waiver-decks-stacked-against-petitioners/

(36) US Chamber of Commerce Project No Project,
http://www.uschamber.com/sites/default/files/reports/PNP_EconomicStudyweb.pdf

(37) *Environmentalists Fight Solar, Wind, Renewable Energy*, August 10, 2012, Investor's Business Daily,
http://news.investors.com/080912-621692-environmentalists-fight-solar-wind-renewable-energy.aspx

(38) *Environmentalists Fight Solar, Wind, Renewable Energy*, August 10, 2012, Investor's Business Daily,
http://news.investors.com/080912-621692-environmentalists-fight-solar-wind-renewable-energy.aspx

(39) Associated Press:
http://www.kulr8.com/news/state/9970866.html;
http://www.macalester.edu/windvisual/valleyinfo.html#sum;

(40) *Environmentalists Fight Solar, Wind, Renewable Energy*, August 10, 2012, Investor's Business Daily,
http://news.investors.com/080912-621692-environmentalists-fight-solar-wind-renewable-energy.aspx

(41) *Lawsuit Against Wind Energy Project Near Steens Mountain Pits Green Groups Against Green Project*, Richard Cockle, May 2, 2012, The Oregonian,
http://www.oregonlive.com/pacific-northwest-news/index.ssf/2012/05/lawsuit_against_wind_energy_pr.html
Another

(42) *Court To Consider Injunction In Kern County Wind Case*, Chris Clark, July 27, 2012, Industrial Wind Action Group, http://www.windaction.org/news/35650

(43) *Chapter Joins Suit against Panoche Valley Solar Plant*, Erin Barrite, July/August, 2011, The Loma Prieta Sierra Club, http://lomaprieta.sierraclub.org/loma-prieta/story/action/chapter-joins-suit-against-panoche-valley-solar-plant/2895

(44) *Environmentalists Fight Solar, Wind, Renewable Energy*, August 10, 2012, Investor's Business Daily, http://news.investors.com/080912-621692-environmentalists-fight-solar-wind-renewable-energy.aspx

(45) *Sierra Club, NRDC Sue Feds To Stop Big California Solar Power Project*, Todd Woody, March 27, 2012, Forbes, http://www.forbes.com/sites/toddwoody/2012/03/27/sierra-club-nrdc-sue-feds-to-stop-big-california-solar-power-project/

(46) *Greens against Green Energy*, Wall Street Journal, September 5, 2012, http://online.wsj.com/article/SB10000872396390443618604577621751892592534.html

(47) Panel: *Green Jobs Company Endorsed by Obama and Biden Squandered $535 Million in Stimulus Money*, John Rossomando, February 22, 2011, Daily Caller.

(48) *More Solar Companies Led By Democratic Donors Received Federal Loan Guarantees*, Daily Caller, September 29, 2011

(49) *Obama's Solar Failures Abound*, March 31, 2012, Investor's Business Daily

Section 4: Regulation Run Amok

(1) Interview with Dr. Jay Lehr, *Defender of Our Industry*, Paradigm and Demographics, March 1, 2011, http://paradigmsanddemographics.blogspot.com/2011/03/interview-with-dr-jay-lehr-defender-of.html

(2) National Center for Policy Analysis bulletins.

(3) *New EPA Air Regs Will Cost Billions of Dollars*, April 11, 2011; *The EPA's New Air Quality Regulations: All Pain, No Gain, Part One*, April 11, 2011

(4) National Center for Policy Analysis bulletin, *New EPA Air Regs Will Cost Billions of Dollars*, April 11, 2011; *The EPA's New Air Quality Regulations: All Pain, No Gain, Part Two*, April 11, 2011.

(5) *Overruled*, Investor's Business Daily, June 3, 2011, *EPA: Jobs Don't Matter*, Investor's Business Daily, April 19, 2011

(6) *Regulatory Agencies Staffing Up*, Investor's Business Daily, August 16, 2011

(7) Marlo Lewis, March 21, 2011, Pajamas Media.Com
http://pajamasmedia.com/blog/epa%2%/80%99s-greenhouse-power-grab-baucus%e2%80%/90s-revenge-democracy%e2%80%/99s-peril/

(8) *The Senate's Showdown*, March 28, 2011, Wall Street Journal

(9) Editorial: *EPA's chilling effect*, Washington Times, January 14, 2014,
http://www.washingtontimes.com/news/2013/jan/4/epa-s-chilling-effect/

(10) *EPA's Cross-State Pollution Rule Upheld By Supreme Court*, Dina Cappiello and Sam Hananel, Huffington Post, April 29, 2014
http://www.huffingtonpost.com/2014/04/29/epa-cross-state-pollution-rule_n_5232875.html

(11) *Frack, Baby, Frack*, Investor's Business Daily, May 10, 2011

(12) *Bakken Shale Promises Big Oil Production*, Donald D. Gold, March 24, 2011, Investor's Business Daily

(13) *The Madness of New York*, Wall Street Journal, December 16, 2010

(14) *Chesapeake Spill Heightens Pressures*, Ben Casselman, April 28, 2011, Wall Street Journal

(15) *The Madness of New York*, Wall Street Journal, December 16, 2010

(16) *EPA Frac Study to Focus on Water Impact*, Maureen N. Moses, July 7, 2010, AAPG Explorer
http://www.apg.org/explorer/2010/07/jul/epafrac0710.cfm

(17) *Fracking for Natural Gas: EPA Hearings Bring Protests*, Mark Clayton, September 13, 2010, Christian Science Monitor
http://www.csmonitor.com/layout/set/print/content/view/print/325494

(18) *Halliburton Develops Eco-Friendly Fracking Fluid*, Brady Nelson, Environment and Climate News, March 2011.

(19) Roy W. Spencer, *Climate Confusion*, (Encounter Books, 2008), 98-99

(20) *GAO Report Exposes Millions in Environmental Litigation Fees for the First Time*, US Senate Committee on Environment and Public Works: Minority Page, August 31, 2011,
http://epw.senate.gov/public/index.cfm?FuseAction=Minority.PressReleases&ContentRecord_id=20ba71b7-802a-23ad-4ca7-a06341934622

(21) *EPA's New Regulatory Front: Regional Haze and Takeover of State Programs*, US Chamber of Commerce,
http://www.uschamber.com/sites/default/files/reports/1207_ETRA_HazeReport_lr.pdf

(22) US Chamber Report Reveals that EPA's Takeover of States' Regional Haze Programs is all Cost, No Benefit, US Chamber of Commerce,
http://www.uschamber.com/press/releases/2012/july/us-chamber-report-reveals-epa%E2%80%99s-takeover-states%E2%80%99-regional-haze-programs-all

(23) Press Release/Letter, *Vitter Warns Louisiana of EPA's Secret 'Sue and Settle' Deals, Could Impact State*, January 22, 2013

(24) *Sen. Vitter Hits the Ground Running*, Myron Ebell, February 5, 2013, Global Warming.com
http://www.globalwarming.org/2013/02/05/sen-vitter-hits-the-ground-running/

(25) Christopher Horner, *The Liberal War on Transparency:*

Confessions of a Freedom of Information 'Criminal', October 2, 2012
http://www.amazon.com/gp/product/1451694881/ref=as_li_
tf_tl?ie=UTF8&camp=1789&creative=9325&creativeASIN=145
1694881&linkCode=as2&tag=chrishorneron-20

(26) *Public Interest Group Sues EPA for FOIA Delays, Claims
Agency Ordered Officials to Ignore Requests*, The Washington
Examiner, January 28, 2013,
http://washingtonexaminer.com/public-interest-group-sues-epa-
for-foia-delays-claims-agency-ordered-officials-to-ignore-
requests/article/2519881

(27) Press Release, The Environmental Law Center of the
American Tradition Institute, January 28, 2013

(28) Marita Noon, Executive Director, Energy Makes
America Great, Inc.

(29) *Why We Support a Revenue-Neutral Carbon Tax*, April 8,
2013, Wall Street Journal,
http://online.wsj.com/article/SB1000142412788732361160457
8396401965799658.html

(30) *Carbon Tax Is Both Pointless and Inflationary*, Steve Milloy,
November 14, 2012, Investor's Business Daily,
http://junkscience.com/2012/11/14/carbon-tax-is-pointless-
and-inflationary/

(31) *Will the Carbon Tax Make a Comeback*, William O'Keefe,
December 21, 2012, Wall Street Journal.

(32) *Why the GOP Will not Support Carbon Taxes (if it wants to
survive)*, Marlo Lewis, November 26, 2012,
http://www.globalwarming.org/2012/11/26/why-the-gop-
will-not-support-carbon-taxes-if-it-wants-to-survive/

(33) *Will the Carbon Tax Make a Comeback?*, William O'Keefe,
December 21, 2012, Wall Street Journal,
http://online.wsj.com/article/SB1000142412788732446930457
8145640617261224.html

Section 5: Truly Scary UN Agendas

(1) Sovereign Independent
http://www.sovereignindependent.com/?p=18097

(2) *Climate Talks or Wealth Redistribution Talks?*, Nicolas Loris, November19, 2010, Heritage.org,
http://blog.heritage.org/2010/11/19/climate-talks-or-wealth-redistribution-talks/

(3) Paul Ehrlich, Anne Ehrlich, and John Holdren, *Ecoscience: Population, Resources, and Environment*, W. H. Freeman and Company, 1977), p. 954.

(4) Paul Ehrlich, *The Machinery of Nature*, Simon & Schuster, New York, 1986, p. 274

(5) Obama's Top Science Adviser to Congress: *Earth Could Be Reaching Global Warming 'Tipping Point' That Would Be Followed by a Dramatic Rise in Sea Level*, December 9, 2009 , CNS News,
http://cnsnews.com/news/article/obamas-top-science-adviser-congress-earth-could-be-reaching-global-warming-tipping#sthash.xXbuEL6b.dpuf

(6) *Green Climate Fund Fundamentally Transforms the Global Economy*, Cathie Adams, April 29/ May 6, 2011, Eagle Forum
http://www.eagleforum.org/un/2011/11-05-06.html

(7) *Climate Change is Not About the Environment, But About Redistributing Wealth*, Conference of the Parties 16, Cathie Adams, Eagle Forum
http://www.eagleforum.org/un/2010/10-12-01.htm

(8) *After a Year of Setbacks, UN Looks to Take Charge of World's Agenda*, George Russell, September 8, 2010, Fox News,
http://www.foxnews.com/world/2010/09/08/years-setbacks-looks-world-leader/print#

(9) *After a Year of Setbacks, UN Looks to Take Charge of World's Agenda*, George Russell, September 8, 2010, Fox News,
http://www.foxnews.com/world/2010/09/08/years-setbacks-looks-world-leader/print#

(10) *The History of the Global Warming Scare*, July 28, 2009,

Global Warming Science
http://www.appinsys.com/GlobalWarming/GW_History.htm
 (11) Investors.com
http://news.investors.com/ArticlePrint.aspx?id=554439
 (12) *More UN Insanity We are Paying For: Isn't it Crazy to Continue?*, Larry Bell, Forbes Opinions, December 6, 2011
 (13) *UN Agreement Should Have All Gun Owners Up in Arms*, Larry Bell, Forbes Opinions, June 7, 2011,
http://www.forbes.com/sites/larrybell/2011/06/07/u-n-agreement-should-have-all-gun-owners-up-in-arms/.
 (14) *An Accountable UN At Long Last?*, September 1, 2011, Investor's Business Daily
http://news.investors.com/Article/583459/201108311840/An-Accountable-UN-At-Long-Last-.htm
 (15) *The UN's Global Warming War on Capitalism: An Important History Lesson*, Larry Bell, Forbes Opinions, January 22, 2013.
http://www.forbes.com/sites/larrybell/2013/01/22/the-u-n-s-global-warming-war-on-capitalism-an-important-history-lesson-1/
 (16) British Researcher: *No Global Warming in 18 Years*, Drew MacKenzie, September 9, 2014, Newsmax,
http://www.Newsmax.com/Newsfront/Global-warming-climate-change-Christopher-Monckton/2014/09/09/id/593471/

Appendix Articles

Prospering, But Pity Those Paramecia!, Larry Bell, Forbes Opinions, April 24, 2012

Appendix 3-1: *Green Energy Promotions Rely Upon Colorblind Public*, Larry Bell, Newsmax, September 8, 2014

Appendix 3-2: *Wind's Overblown Prospects*, Larry Bell, Forbes Opinions, March 2011

Appendix 3-3: *Green Power Gridlock: Why Renewable Energy Is No Alternative*, Larry Bell, Forbes Opinions, December 10, 2013

Appendix 3-4: *Ten Reasons to Care That E15 Ethanol is on the Way to Your Gas Station*, Larry Bell, Forbes Opinions, September 23, 2012

Appendix 3-5: *Environmental Groups Strongly Endorse 'None of the Above' Energy Plans*, Larry Bell, Forbes Opinions, March 12, 2013

Appendix 4-1: *EPA and Enron End-Runs of Congress: Lisa Jackson Serves US Industry a Thanksgiving Turkey*, Larry Bell, Forbes Opinions, December 2010

Appendix 4-2: *New Congress Must Rein In Runaway EPA*, Larry Bell, Newsmax, June 2, 2014

Appendix 4-3: *EPA's Insanely Ambitious Agenda If Obama Is Reelected*, Larry Bell, Forbes Opinions, November 4, 2012

Appendix 4-4: *EPA Mandates that New Coal Plants Prevent Nonexistent Climate Problem with Unavailable Solution*, Larry Bell, Forbes Opinions, October 1, 2013

Appendix 4-5: *EPA's Secret and Costly 'Sue and Settle' Collusion with Environmental Organizations*, Larry Bell, Forbes Opinions, February 17, 2013

Appendix 4-6: *EPA Head Admits Being Clueless About Any Obama Climate Plan Benefits*, Larry Bell, Forbes Opinions, September 22, 2013

Appendix 4-7: *Carbon Tax...Are Republicans Really That Stupid?*, Larry Bell, Forbes Opinions, April 16, 2013

Appendix 5-1: *Yes! We Should Defund The UN's Intergovernmental Panel On Climate Change!*, Larry Bell, Forbes

Opinions, February 24, 2013

Appendix 5-2: *Books: Robert Zubrin's Merchants of Despair Reveals Racism and Genocide Cloaked In Green Camouflage*, Larry Bell, Forbes Opinions, July 31, 2013.

About the Author

LARRY BELL IS an endowed professor at the University of Houston where he founded the Sasakawa International Center for Space Architecture (SICSA) and the graduate program in space architecture. He is the author of *Climate of Corruption: Politics and Power Behind the Global Warming Hoax*, *Cosmic Musings: Contemplating Life Beyond Self* and more than 300 online articles on a wide variety of topics as a *Forbes* and *Newsmax* Contributor, some of which have also been featured in their hard-copy magazine publications.

Larry's professional aerospace work and interviews have appeared in numerous TV and print media productions which include the *History Channel*, *Discovery Channel-Canada*, *NASA Select* and leading national and international newspapers, popular magazines and professional journals. His many awards include certificates of appreciation from NASA Headquarters and two highest honors for his contributions to international space development awarded by Russia's leading aerospace society.

CPSIA information can be obtained at www.ICGtesting.com
Printed in the USA
LVOW11s1107120616

492276LV00007B/587/P